W9-ATF-629

738.6 Hilliard, Elizabeth,
Hil Designing with
c.1 tiles
39.95

MY '96

Mount Laurel Library
Mount Laurel, NJ 08054
Phone 234-7319

DEMCO

with TILES

DESIGNING

with TILES

Elizabeth Hilliard

ABBEVILLE PRESS PUBLISHERS
NEW YORK LONDON PARIS

B V L 39.95
87957

First published in the United States of America in 1993
by Abbeville Press, 488 Madison Avenue, New York, NY 10022
First published in Great Britain in 1993 by Pavilion Books

Text copyright © Elizabeth Hilliard 1993

Designed by David Fordham

All rights reserved under international copyright conventions.
No part of this publication may be reproduced in any form or by any means, electronic or mechanical, including photocopying, recording, or by any information storage and retrieval system, without permission in writing from the publisher. Inquiries should be addressed in writing to Abbeville Publishing Group, 488 Madison Avenue,
New York, NY 10022.

ISBN 1-55859-639-9

Printed and bound in Hong Kong. Produced by Mandarin Offset.

A CIP catalog record for this book is available from the British Library.

First American Edition, Second Printing.

Jacket, front: An interior in the Low Countries (see page 41)
Jacket, back, clockwise from upper left: a tiled fireplace surround (see page 104); a checkerboard kitchen floor (see page 96); color accents in a bathroom (see page 94); a French country house floor (see page 80).

CONTENTS

CHAPTER ONE

The ISLAMIC WORLD

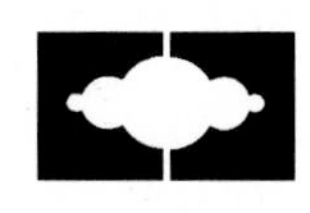

THE STORY OF TILES IN THE ANCIENT AND ISLAMIC WORLDS DOES NOT FOLLOW A neat straight line in time and place. Instead, it is as colourful and intricate as the design on an elaborate Islamic tile, leaping about from country to country and from century to century. The ancient Egyptians faced buildings with bricks coated with coloured glaze, and the Abyssinians decorated monuments with relief animals formed from glazed tiles. In some periods the tile's popularity wanes, and with it the whole state of the industry, while in others the inventiveness of designers and artisans and the imagination of patrons and architects lift the slab of baked earth to heights of brilliance. At all times, however, the tile depends on a peaceful and stable political atmosphere of the type in which all arts and crafts have the best chance of thriving; in which the attention of rulers is focused not on fighting wars but on glorifying themselves and their gods by embellishing palaces and places of worship.

The most brilliant moment in the tile-making industry came in Turkey in the middle of the sixteenth century, under the rule of the sultans of the Ottoman Empire and in particular Sultan Suleiman the Magnificent. Many of the buildings created during this period can still be seen today, monuments both to patrons and to craftsmen. The sultans were enthusiastic patrons, and their programme of building magnificent mosques and palaces coincided with a steady improvement in the tile makers' technical and artistic achievements. Tile-making was part of a wider ceramics industry; sometimes tiles were eclipsed by utility ceramics – dishes, bowls, jugs and other vessels for practical use – but the sixteenth century was not one of those periods.

Like other artisans, tile makers were an itinerant breed, moving between cities and countries according to the opportunities for work offered by each place. The two centres of the tile-making industry were outside Istanbul itself, in Iznik and Kutahya. Designs were provided by artists at court and then sent out to Iznik and Kutahya, where samples were made. In the case of at least one commission it is known that the two towns were in competition with each other; Iznik won.

As far as we know, the tiles from Iznik and Kutahya were made from local clay of excellent quality which was quarried at or near the factories, and fired in kilns

OPPOSITE *Tiles in the Alhambra, a group of Moorish royal palaces and fortifications linked by gardens, courtyards and terraces, on a hilltop overlooking Granada, in Andalucia in Spain. Groups of four square green tiles are surrounded by black tiles and divided by white panels and eight-pointed stars of random colour.*

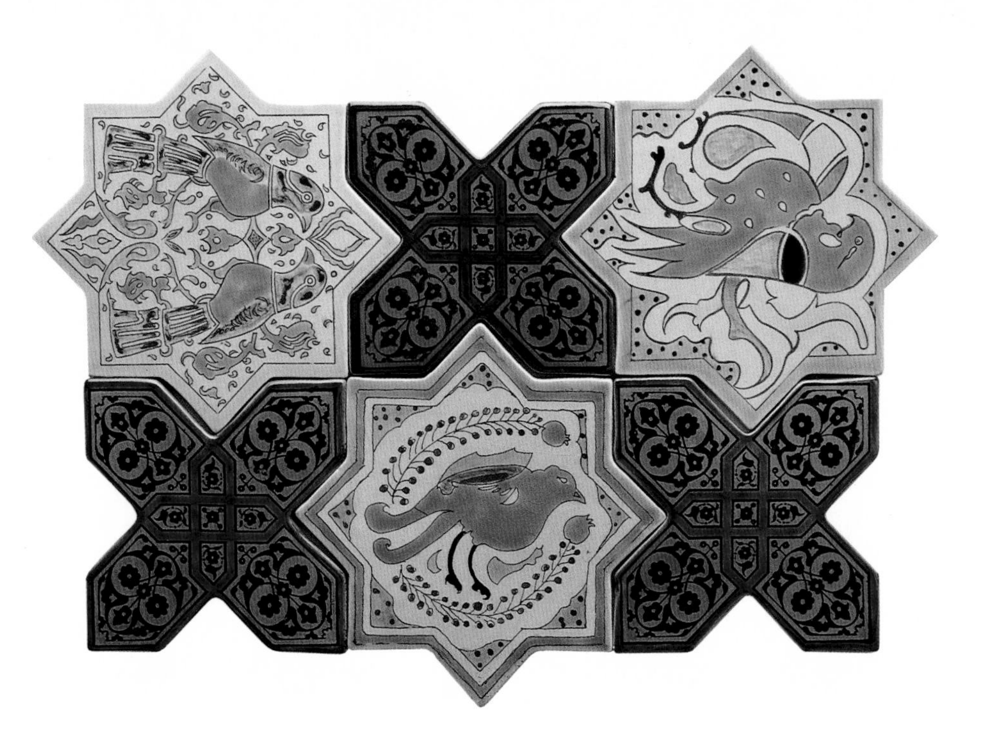

RIGHT *Modern tiles made in Turkey to the traditional Islamic design of eight-pointed stars separated by pointed crosses. The stars are decorated with exotic birds and stylized foliage; the crosses enclose lattice shapes and abstract flower forms. The darker colours of the crosses throw into relief the delicate birds, decorated with turquoise and dark blue glazes.*

powered by the burning of timber, of which there was plenty in the immediate countryside. Today, kilns and other elements of mechanization may be powered by electricity or modern generators, yet access to the raw material, clay, is still an important consideration in the siting of tile factories. All over the world, tile-making factories dig the clay from surface pits in their own back yards, transporting it only a few hundred yards and processing it directly, or after minimal blending, without complicated purification techniques. Clay could be called nature's own manufacturing material.

The clay of Iznik was white when fired and offered an ideal background for a brilliant white underglaze, against which coloured decoration stood out especially well. In the fourteenth century, increased trade with China brought Chinese ceramics into the country and made blue and white decoration fashionable, a combination which suited Iznik products.

The white glaze achieved by the Iznik tilers was of marvellous quality, and did not crackle when fired. Black or green and blue decoration on white was popular, as was black under a limpid turquoise glaze. Shapes of tiles included the predictable square and other, more elaborate shapes, inherited from previous periods of tile-making, which we think of today as being typically Islamic: the hexagon, sometimes with plain-coloured triangles between, and the eight-pointed star showing flowers and birds, which was used in combination with a geometrically-decorated pointed cross. Stencils were used to enable the craftsmen to reproduce designs accurately.

From the town of Iznik, finished work was transported to Istanbul by first being taken overland on horseback to the port of Karamorsel, and thence by boat to the capital. The total time between manufacturing the tiles and seeing them installed on a building was only a matter of weeks.

Tiles were used on the outside and inside of buildings, covering whole walls, façades and domes, not merely in decorative panels placed here and there. The over-

ABOVE AND LEFT *Two arrangements of tiles of different shapes which appear frequently on Islamic buildings decorated with tiles. The hexagons enclose lively flower forms and are separated by triangles of glowing dark blue. They can also be laid without the triangles, in a honeycomb pattern.*

all effect is stunning. Outdoors, in brilliant sunshine, tiled buildings radiate colour, the white and blue glazes reflecting light in a way which would outshine even a building covered in real turquoise. Decoration on the tiles was conceived overall as well as in detail, so that there is as much to look at and reflect upon at a distance as there is on close inspection. Up close, you can read the inscriptions, enshrined in particular panels, or leading the eye around the building, and see in detail the lively portrayal of trees and flowers. From a distance, the eye follows the decoration upwards through intricate arabesques to the higher reaches where individual images blend into the soaring curves of the architecture.

Besides two blues – a vivid turquoise and a deeper shade similar to indigo – colours on Iznik tiles include green and black. But the colour that above all others characterizes Iznik tiles at their pinnacle of artistic achievement is red. Variously described since then as 'radiant tomato-red' or 'sealing-wax red', the colour was a technical breakthrough which demonstrated the skill of the tile makers and gave the designers much greater scope. Until now, the only achievable red was more like a shade of brown, after firing, due to chemical reactions in the glaze. The tile makers of Iznik solved the problem by producing first a slightly mottled version, then a pure brilliant red, which they applied thickly for maximum effect. Not only was the colour splendid in its own right, but it made a vivid contrast to existing blues and greens, and in particular to another new colour, a viridian green. Manganese, a brownish-purple colour, was in most cases dropped from the palette, because it was dull by comparison with the new red and because it did not contrast well with it.

Designers made the most of the opportunities offered by the range of colours now available. Reflecting the Turkish tastes of the period, tiles were decorated with swirling patterns, often incorporating realistic portrayals of peonies, pomegranate flowers, roses, hyacinths, violets, cypress trees, vines, birds and even imitations of marble. Many of the motifs also appeared in contemporary carpets and textiles and other products of the decorative and applied arts. This homogeneity was not surprising, since the same court artists were supplying designs for these and for the tiles of Iznik and Kutahya.

It was not only the grand palaces and mosques, designed as a reflection of earthly and spiritual power and glory, that were decorated with tiles. Domestic quarters such as the harem were clad in tiles, as were fountains, libraries, courtyards and the more wealthy homes. The Sultan is even reputed to have decorated one of his barges with tiles. Tiles were expensive and their use a reflection of status, but they were available to any wealthy person who could pay for them. They were not simply confined to walls, although this is where they were most used. Window openings, for instance, were given ceramic surrounds and even filled with open-work lattice tiles.

The burgeoning success of the Islamic tile industry could not be maintained indefinitely. One of the signals that all was not well occurred in the middle of the seventeenth century, with a change in tone of the red and other colour glazes. The red lost

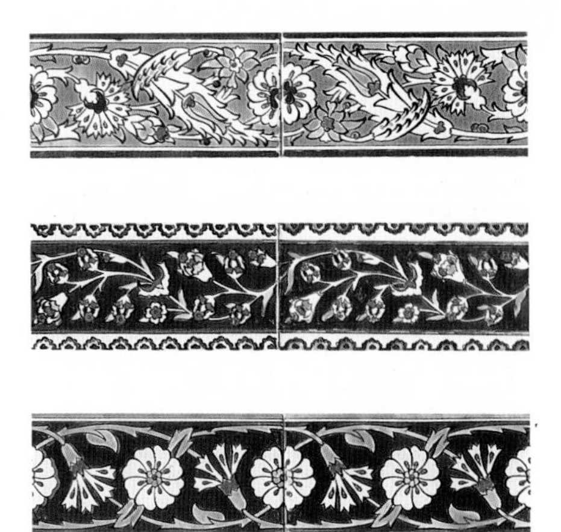

ABOVE *Modern tile borders of Iznik design which can be used to finish a dado of Turkish tiles, outline a door or window, edge a tile-topped table or frame a mirror.*

OPPOSITE *Modern Iznik tiles with the luscious 'sealing-wax red' which was one of the great achievements of Islamic tile production. In times of relative peace, the sultans of the Ottoman Empire turned their attention to embellishments for their mosques and palaces.*

ABOVE *Modern reproduction of a tile panel from the Selimiye mosque at Edirne. This city was the sultans' capital for a period and is the site of some of their most beautiful buildings.*

OPPOSITE *A modern interior in North Africa, with a fountain providing the focus in a covered courtyard. Pool, walls and pillars are faced with glazed tiles with contemporary interpretations of Islamic designs. The floor is covered with marble tiles.*

its vigour and became dull and brownish, while other colours also became muted and muddied. The quality of glazes deteriorated and designs also lost their edge, becoming sketchier and more monotonous and repetitive. One of the causes of this deterioration was the weakening of court patronage. The Turkish Empire was no longer rock solid and the sultans' attention turned from their surroundings in Istanbul to the wider world where their power was under threat.

By 1648 only nine tile-making factories remained, whereas in around 1600 there had been approximately three hundred. Iznik tile makers did not sit by idly and watch their industry decline; they sought other markets, and gained commissions from further afield, from countries such as Greece. But the lack of sufficiently big local and foreign orders and the deterioration in the quality of the tiles were problems too great to overcome. By the end of the seventeenth century the formerly great tile industry in Iznik had almost faded away. It did not vanish completely, however. Enough craftsmen remained for an attempt to be made to revive the industry, not in Iznik but in Istanbul, in the early eighteenth century. The attempt was not a great success. It proved impossible to recapture the combination of vivid colours on a pure ground with a flawlessly clear overglaze, and the introduction of new colours, such as a lurid yellow, seemed to detract rather than add to the effect.

An even more significant difference in these eighteenth-century tiles was the approach to design. One of the elements that made earlier tile-covered buildings so magnificent, going back to the Assyrians and Persians many centuries before, was the way in which the decoration was conceived as a whole, over the entire building. Eighteenth-century designers tended to focus on individual tiles, which resulted in cruder, repetitive patterns rather than the great dramatic swirls and schemes that romped across the monuments of previous ages.

The Turkish people did not lose their love of decorating with tiles during this period, and are known to have imported quantities of blue and white tiles from Holland in the middle of the eighteenth century. Another development in the same century gave a bizarre twist to the history of the Islamic tile: it was hijacked, so to speak, by Christians, and in particular by the wealthy Armenian Christians who had fled to Turkey and the eastern Mediterranean to escape the upheavals of war in their own country. When they returned to Armenia, a period of greater religious freedom enabled them to give generously to their churches, and splendid church interiors came into being, created with tiles from Kutahya.

In the twentieth century, tile-making in Kutahya has once again experienced a resurgence, based on traditional Iznik and Islamic designs and on commissions from the authorities involved with repairing and rejuvenating Jerusalem after the First World War. Public and religious buildings, and on a smaller scale private homes, have continued to be covered with large areas of tiling, which are not considered particularly expensive and cannot be surpassed for decorative effect and practicality. From time to time, special, prestigious commissions, like the tiles made for the Hilton hotel in Istanbul

in the mid-1950s and the recently built Makkah Gate Bridge, give the industry a welcome boost.

In the history of tiles, no period is totally self-contained. Ideas and designs leapfrog across the centuries and from one culture to another, echoing the tile maker's roving quest to find work through the ages. Lustreware, for example, on which glazes have a metallic ingredient and finish, was one of the most exciting products of the Islamic ceramics industry. Lustre, together with the vibrant colours and swirling vigour of the designs on the finest Islamic tiles, inspired an Englishman, Edward De Morgan, to design and make some of the most beautiful and distinctive tiles and other ceramics of the later nineteenth century. And in 1921, an Armenian factory in Kutahya made decorated 'Islamic' tiles for an inlaid sideboard by the Arts and Crafts designer C. R. Ashbee, demonstrating a continuing awareness of and interest in eastern tiles on the part of western artists.

Today, Islamic tiles are made in Turkey in a small but flourishing industry, still craft-based with only a minimal amount of mechanization in the production process, and apparently depending on the enthusiasm and commitment of individual industrialists. Over the centuries, meanwhile, Islamic religion and culture spread across vast areas of Europe, Africa, the East and Asia. It even progressed as far north-west as Spain, where it was taken up by the Moors. In Granada, southern Spain, the Alhambra is a dazzling monument to the effect to which tiles can be put when applied *en masse* to a building, especially when combined with water. Today in Granada, a small number of factories still produce tiles in the colours and designs of the fourteenth century. As always, wherever they are made the quality of modern Islamic tiles depends as much on design as on technical standards, and studies by modern designers of the best achievements of historical tiles have led to new tiles in satisfyingly clean vigorous patterns which do no discredit to the great tradition.

ABOVE *Two more examples of modern Iznik tiles based on traditional Islamic designs.*

OPPOSITE *The Arab Hall at Leighton House, built in London in 1866 for the artist Frederick Lord Leighton, decorated with antique Islamic tiles and a mosaic frieze by Walter Crane.*

CHAPTER TWO

MEDIEVAL TILES

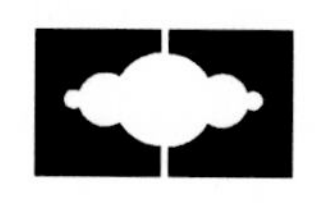

IN SHIFTING OUR ATTENTION FROM THE ISLAMIC WORLD TO MEDIEVAL ENGLAND, our focus also moves, from walls down to floors. Many of the Arabic and other Islamic peoples were nomadic, packing up their treasured possessions and taking them with them as they travelled from place to place. One of the great traditions of the Islamic world lay in making fine carpets and rugs, which were used to cover floors and could if necessary be rolled up with ease and removed to the next town or camp.

By contrast, England had been an important outpost of the Roman Empire, whose technique for embellishing floors in their palaces and grand villas was the mosaic. There are strong parallels between tiled floors and mosaic floors, although the materials involved are different. Roman mosaics were constructed from tiny squares of stone in their natural colour and occasionally from coloured glass, rather than from the man-made fired components that make up a tiled floor. But the reasons for laying mosaics were similar to those behind tiled floors: they were decorative and prestigious, hard-wearing and practical – qualities which are equally valid today.

With mosaics, the small size of the pieces, and the elaborate and figurative subjects they showed, meant that the best effect was achieved by laying the pieces to follow the lines of the subject, be it a mythological god or an abstract decorative motif. By contrast, medieval tiled floors used larger pieces in whole geometric shapes or squares, and the layout was correspondingly more regular.

The Romans also occasionally made floors from stone cut into shapes, rather like pieces of tile mosaic. This was called *opus sectile*. One of the few examples of this type of floor to survive in Britain can be found at the Roman palace in Fishbourne. The Romans did make tiles, but they were unglazed, utilitarian building components for roofs, wall friezes and floors. These were undecorated, and did not have the prestige either of Roman mosaic or of decorative tile pavements in medieval England.

A few Roman floor mosaics had been uncovered and were known to people in medieval England, but the source of the enthusiasm for tiled floors was more obviously contemporary. Small quantities of Spanish tiles had come into the country with

ABOVE *These borders are a modern example of mosaic, made from many pieces of coloured stone. The Romans occupying Britain used mosaic for floor decoration.*

OPPOSITE *Tile mosaic was a popular form of floor covering in medieval English abbeys. This display in the British Museum in London shows six exuberant patterns and several border designs. Metal templates were used to cut clay into shaped tiles before firing. After glazing the pieces were laid in place.*

RIGHT *Mosaic tiles are made by arranging the pieces of coloured stone face down in the required pattern in a frame, then fixing these in place with a mortar. Today the mortar is a cement-based compound. The process is as labour-intensive today as it was centuries ago, with hundreds of pieces of stone in each tile.*

travellers and tradesmen, and returning crusaders brought with them a knowledge of and interest in eastern tiles. English craftsmen were already sufficiently skilled in making and using lead glazes, and were able to apply this expertise to the making of floor tiles.

The deciding factor in the flowering of medieval tile art in England was probably the church. Gradually, over the centuries, it had reached an unrivalled position of power and influence, based to a great extent on the wealth that came with landowning on a large scale. Abbots and bishops became patrons of the arts, commissioning buildings which reflected the extent of their secular power as well as glorifying God. Architects and craftsmen looked as much to these churchmen for work as they did to kings and princes. Accordingly, it was in cathedrals and abbeys that many of the most magnificent tiled floors were to be found.

In the Cistercian religious settlements of England, the lay brothers and the monks themselves were a source of unpaid labour in the tile industry. Among other duties, these men would lay the floors and even help make the tiles in the tilery, which specialist tile makers would construct on-site for the purpose of fulfilling a large commission. In the first decades of the thirteenth century, when the enthusiasm for tiled floors was first apparent in England, this is largely how the tiles were manufactured. It was a system which took a good part of the year to complete. Clay was dug out of the ground in the autumn months, turned in winter, and worked into tiles in spring. In the meantime, the tile-making workshops and kilns were being prepared. Tile blanks were cut, then dried outdoors or in a shed kept warm with a fire, before being decorated, trimmed, glazed and fired.

In 1347 large numbers of people died in a plague which swept across England and was known as the Black Death. This had many results, including a diminution in the numbers of men in the workforce and a general impoverishment of the community, patrons and workmen alike. As a consequence, the way in which tiles were manufactured underwent a change. Commercial tileries had already existed for nearly a century, satisfying the increasing demand for tiles from people such as successful merchants, who wanted tiled floors in their most important rooms but could not afford to set up their own tilery. Following the Black Death, on-site tileries virtually ceased to exist, while commercial tileries became the normal source of supply and charged customers for the transport of their tiles as well as for the tiles themselves.

The most immediately striking difference between medieval English tiles and the greatest achievements of the Islamic tile-making industry is their colour. In place of

LEFT *Sixteenth-century tiles from the manor at Southam de la Bere in Gloucestershire. Their heraldic decorations retain their freshness today. The TS in the tile at upper left refers to Abbot Thomas Segar. These tiles are decorated with liquid clay, called 'slip', which has been poured into the indentations made by pressing a wooden stamp into soft tiles.*

ABOVE *Incredibly, these beautiful relief and counter-relief tiles from the second half of the fourteenth century are discards, found on the waste heaps around the commercial tilery at Bawsey, near King's Lynn in Norfolk.*

cool brilliant whites, blues and greens, English medieval tiles offer warm, earthy tones of reddish-brown, clotted cream, various shades of green and a sooty black. The different colours evoke a dramatic change in atmosphere: Islamic tiles speak of hot sunlight glancing off dazzling blue domes and making patterns in the sparkling water of pools and fountains; English medieval tiles conjure up images of the mysterious dusky gloom of soaring Gothic churches, and the crackle and glow of fires indoors to keep at bay the weather and wintry landscape outside. The same contrast is true in today's homes: Islamic patterned tiles hint at sunshine and eastern sophistication; stoneware tiles in earthy colours suggest warmth and wellington boots.

The colours of medieval tiles were the result of the lead glazes that were used in their decoration. The basic glaze turned white clay yellow, due to iron impurities in its make-up. Other metals could be added to create other colours: extra iron for a dark brown, small amounts of copper or brass for green and larger amounts for black. The final colour of a decorated tile also depended on the colour of the clay from which the tile was made and with which its pattern, if any, was formed. The basic glaze, for example, looked yellow on white clay, but olive-green on grey, and brown on red.

There are two elements in the decorative impact of medieval tiled floors: the shape and arrangement of the tiles, and their surface decoration. The simplest floors had square, undecorated tiles, and the next simplest had square tiles, each decorated with one of two coloured glazes, which were arranged alternately in a chequerboard pattern. These became popular in the middle of the sixteenth century, right at the end of the greatest period of medieval tile art. Floors decorated in this way were part of the new interest in Italianate architecture during the sixteenth century, and contributed to the disappearance of the decorated tile.

The most dramatic floors to make an impact through the shape and arrangement of the tiles were the so-called 'tile mosaics'. These were formed from tiles which were undecorated except for a coloured lead glaze. The pieces of the tile mosaic were cut into their eventual geometric shapes before being fired, and later fitted together like a jigsaw to form an uninterrupted floor. The craftsman, monk or lay brother who laid the floor would start in the middle, so that any irregularity in the size or shape of the building could be accommodated in the borders around the edges.

One such floor which survives today was from Byland Abbey in Yorkshire, now on display in the British Museum. From a sun-like flower in the centre, rings of geometric pattern radiate out until they nearly reach the edges of the floor. The patterns within the rings are bursting with vitality, though some consist of little more than simple diamonds or diagonal squares. The main band of pattern is a necklace of wheels, each with curved lozenge-shaped spokes. Each corner of the square floor, outside these concentric circles, is further decorated with an individual ringed star and two moons.

Fragments of other mid-thirteenth-century floors, one of which comes from another Yorkshire abbey called Rievaulx, further demonstrate the ingenuity of the designers of tile mosaics. One is an extremely simple but effective pattern formed entirely from dark triangles and pale diamonds; another has circles within circles connected by S-shaped curved strips; yet another has an elaborate layout of larger circles punctuated with smaller ones and containing eight-petalled flower shapes. In all the examples, the designer has made maximum impact with a limited range of colours created by lead glaze over the natural clay. Even more striking is the liveliness of the designs, which communicate a real joy in inventive use of interlocking shapes and bold

ABOVE *The vigorous design of this tile mosaic floor from Byland Abbey in Yorkshire, with its rings of pattern radiating out like ripples on water, looks as fresh and modern today as it did when the monks first saw it in medieval England.*

patterns, and speak to us across the hundreds of years since they were created. In this sense they are totally modern.

In thirteenth-century England there was one outstanding example of real *opus sectile* or 'cut work' (the method of decorating floors by covering them with intricate patterns of cut-out shapes in stone that had been popular with the Greeks and Romans). It can still be seen today in Westminster Abbey, where King Henry III and his Abbot of Westminster, Richard de Ware, initiated the magnificent scheme of decoration completed in 1268. The King's ambassador had seen such work in Italy and been deeply impressed – so much so that de Ware was able to bring over from Rome a designer called Odoricus, a talented member of one of the highly skilled families who specialized in stone inlaid floors and who were called the Cosmati. They gave their name to the medieval version of *opus sectile*, which came to be known as 'cosmati work'. Much of the marble used to decorate Westminster Abbey was imported from Italy specially for the purpose. The magnificence and intricacy of this floor inspired other abbeys to attempt their own, tiled, versions.

Making and assembling tile mosaics was an exceptionally tricky and time-consuming method of covering floors, however, and by the middle of the fourteenth century the fashion had passed and no more were made. Two types of floor decoration which remained popular were two-colour tiles, and relief and counter-relief tiles. The latter were mostly made by stamping the pattern into the tiles, either so that the decoration stood out (relief), or so that it was set back from the surface (counter-relief). In both cases, the person carving the wooden block used to stamp the tiles had to remember to reverse the pattern, writing or heraldic motif on his carving, so that it would come out the right way round on the tile. Many tiles of the period show that this important detail could easily slip the memory of the craftsman concerned. Later, the method of stamping pattern on to tiles was used to make line patterns and pictures, and some tiles were simply incised individually by hand to achieve the same effect.

The patterned tiles that achieved greatest popularity were those with two-colour decoration. They too were stamped with a pattern, which was then filled with clay of a contrastingly pale colour and trimmed. Later, instead of clay in a solid state, the indentations were filled with 'slip', watered-down clay which was thickly runny. This was allowed to dry before firing. It was floors decorated with tiles like these that the Victorians uncovered and copied with such enthusiasm in the nineteenth century.

Inlaid two-colour decorated tiles from this period showed a wide range of pictures of people, animals, birds, fish, heraldic symbols, mythical beasts, legendary characters, castles, plants and leaves, portrayed in a simple but lively and spirited manner. Others had abstract decoration in the form of energetically curling swirls and arabesques, and others still showed names, or other references to important church and civic officials. Decorated tiles were often laid in square or rectangular panels across a floor, divided by bands of plain tiles of a contrasting colour. This way, the

OPPOSITE *The Great Pavement of Westminster Abbey is a work of* opus sectile*, made using pieces of stone cut to fit a complex pattern. The designer Odoricus, a member of a leading family of marble workers in Rome called the Cosmati, was involved in creating it, and it was finished in 1268 in the reign of Henry III. Interestingly, it includes some 'recycled' purple porphyry stone known to have been cut from Egyptian mines which were closed in the fifth century* BC.

LEFT *Four tiles make up this royal coat of arms found on the magnificent floor of the Chapter House at Westminster Abbey. In 1258 King Henry III ordered tiles left over from this project to be used up elsewhere, so the floor must have been finished by then. The Victorians restored many medieval tiled floors, including this one.*

patterned tiles, which were more expensive, went further, and their impact was also increased by being framed.

Other floors were designed with an overall pattern, perhaps radiating out in circles like the famous floor for the chapel of Clarendon Palace, built for Henry III, or the particularly beautiful and delicately patterned floor from Muchelney Abbey in Somerset which the Victorians saved and resited in the parish church there. A great many inlaid tiled floors of the medieval period can still be seen today, often surrounded by the ruins of the abbey buildings they adorned. If a reminder were needed, they would bear witness to the longevity both of this type of decoration, set into the tiles, and of tiles themselves.

OPPOSITE *These tiles are from the earliest known floor made with two-colour decoration, at Clarendon Palace in Wiltshire. This style of embellishment came from France and was the most popular tile decoration for nearly three centuries. In 1240 a new chapel was begun for Henry III, with this splendid circular design on the floor.*

CHAPTER THREE

MAIOLICA

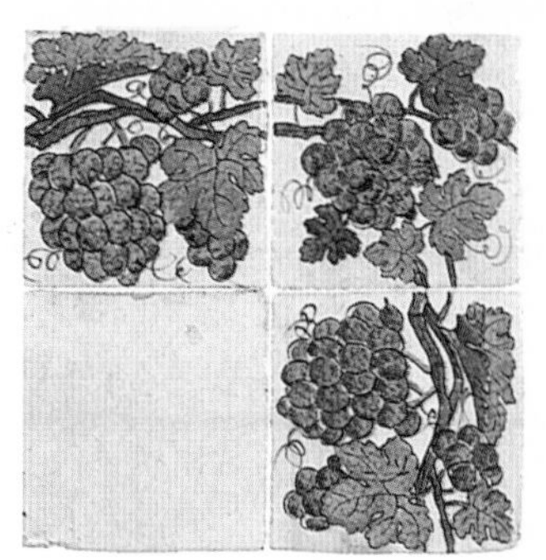

IT IS EASY TO SEE AT A GLANCE THE DIFFERENCE BETWEEN MEDIEVAL TILES AND maiolica: a medieval tile has a rich earthy colour and organic quality; a maiolica tile sparkles with brightly-coloured painted decoration in blue and yellow and green on a white background, and looks sophisticated and delicate. The two tiles seem worlds apart, but in fact the only thing that divides them, in the first instance, is a handful of ashes. It is the addition of oxide (ashes) of tin to the lead glaze that performs the miraculous transformation of a clay-coloured tile into an opaquely white tile. The white glaze has a double effect: it conceals completely the colour of the clay beneath; and it provides a clean white ground, like a sheet of paper, on which the craftsman can paint and which will show off his colours to sparkling effect.

The umbrella name for tiles and ceramics decorated in this fashion is 'tin-glazed ware'. Maiolica is the name given to the wares made in the southern Mediterranean countries – Italy, Spain and Portugal – while in France it is known as faience and in Germany, Fayence. The famous tin-glazed ware made much later in the Netherlands and later still in England is called delftware. The latter has special characteristics of its own, as the next chapter will show, but technically the decorative method is the same.

The basic square tile is made from clay and given a first, 'biscuit', firing. The resulting tile is dry and porous and readily soaks up the liquid tin glaze, into which it is dipped or which is poured over it, leaving a good layer on the surface. (The tile can be dipped in water first so that less glaze is absorbed and more lies on the surface.) The decorator can then paint his patterns and pictures on to the tile, using a variety of coloured glazes formed by the selective addition of yet other metals.

These metallic pigments look unexciting at this stage – dull, matt shades of grey and brown. Known as 'high-temperature colours', they are subjected to a hot glaze firing, which only they can withstand, and become vibrant shades of blue (from the addition of cobalt to the glaze), yellow (antimony), red (iron), green (copper) and purple (manganese). In the eighteenth century, the range of coloured decoration on tin-glazed ceramics was increased by adding, after the high temperature firing, a

OPPOSITE *Modern maiolica tiles with an elegant pattern based on a traditional Italian design. These tiles are made today by centuries-old methods, and though they are now intended for use on walls, they are still made as thick as floor tiles.*

ABOVE *Spain exported its ceramics industry to Mexico when it colonized that country, and the most popular form of decoration was inevitably coloured tin glazes on a milky white background. These modern tiles are decorated with flowers and leaves in patterns which multiply when an area of wall is covered with one design.*

second layer of painted pattern using enamel colours which need firing at a lower temperature. Lustre and gold decoration could also be added in a similar, second, lower temperature firing.

The origins of tin glaze go back to ancient history. From the Near East and, later, Mesopotamia, it spread to Spain and the rest of Europe via North Africa and the Islamic Empire. It also spread to Mexico, exported there by Spanish potters when it became a Spanish colony in the sixteenth century. Islamic Moors and Christians coexisted in Spain more or less peacefully for centuries until the Moors' expulsion in 1610, by which time the two cultures had become intermixed. One of the beneficiaries of this was the Spanish ceramic industry, which gained not only the knowledge of Islamic design, tin-glazing and lustre but also substantial commissions for tiles upon which to use these skills.

From at least the early years of the eleventh century, tin-glazed wares were made in southern Spain. By the thirteenth century, when the startling new blues made possible by the addition of cobalt to glaze were introduced, Malaga was established as an important centre for production. Ceramics were exported from here to countries across Europe. In the later part of the fourteenth century, the Malaga industry was given the commission that above all others has probably made Spanish tile work famous – the Alhambra in Granada.

LEFT *A tile-clad archway in the Alhambra in Granada, southern Spain. Tin glazes were used in southern Spain from the early eleventh century. By the thirteenth century, Malaga had become an important centre for ceramic and tile production and supplied the tin-glazed tiles for the palaces of the Alhambra.*

ABOVE *More designs from the repertoire of modern Mexican tile factories, where craftspeople continue the tradition of decorating tin-glazed tiles by hand.*

Set high on a hill, the Alhambra is a palace of light with a magical atmosphere, even today when it is inhabited mainly by crowds of tourists. Some of the tile work has been cleverly restored, but much is original. Room after room has tiled walls, dados, archways framing magnificent views out over the hills of Andalucia or vistas within the buildings. Courtyards with pools and fountains sparkle with light reflected off tiled surrounds, pavements and pillars. In the brilliant sunshine and heat, the effect is not cold and hard but embracingly cool and serene. Visitors linger and seem eventually to drag themselves away only with great effort. The interior of the Alhambra is a unique example of how beautiful and even restful a tiled interior can be.

By the fifteenth century, Valencia had overtaken Malaga as the most successful centre of the ceramics industry as a whole, while other towns in Aragon and Catalonia led the way in tile-making. Wares from Valencia were especially popular in Italy, travelling there via the Spanish island of Majorca. This is how the name 'maiolica' came to be adopted, even though the style did not originate on Majorca and the island never had a significant practical association with tile production.

One of the most interesting craftsmen making maiolica tiles in Spain was an Italian called Francisco Niculoso who moved to Seville at the end of the fifteenth century and there introduced fine Italian glazes. His tile panels of religious subjects incorporated architectural settings drawn from classical antiquity, and his draughtsmanship was crisp and assured. He portrayed different textures with graphic patterns such as crosshatching and stippling, drawn in minute detail, and he built up a great variety of vivid and subtle colours by layering one coloured glaze over another. Not surprisingly, his work was acclaimed and popular, and he won many prestigious commissions for altar frontals and panels for other prominent positions in the city's religious and civic buildings.

Tile panels of this period varied from small compositions of six or eight tiles to magnificent walls constructed from several thousand. The exact area of the wall was measured out on the floor of the workshop, allowing for doors and windows and any other architectural features. Biscuit-fired tiles coated with plain white tin glaze were then laid on to this space and decorated by the master tile maker. They were numbered or carefully recorded on a plan of allocation before being removed for firing. Once the tiles were made, the huge mural was usually installed by a mason who attached them to the wall with plaster in the correct formation.

Spain's neighbour Portugal also had a rich tradition of tin-glazed decorated wall tiles. In the course of the fifteenth and sixteenth centuries, small tiles made in geometric shapes and known as *alicatados* were superseded by larger tiles which made covering whole walls a quicker and technically easier task. These tiles began to be produced in Lisbon in the second half of the sixteenth century, and were given the name *azulejos*, from the Arab word for ceramic mosaic, *zuleija*. Thus began one of Europe's most outstanding tile traditions.

For centuries, Portuguese architecture had gained much of its effect from applied decoration, using decorative painting, gilding, wood-carving and inlaid marbles to embellish the surfaces of buildings, both inside and out. The discovery of the potential of tiled surfaces gave Portuguese craftsmen a new and refreshing outlet for their talents. They developed a wonderfully free and creative style, painting panels with vigorous draughtsmanship and incorporating designs and images from a mass of sources. These panels covered whole walls, even whole rooms, and were extensively used outdoors on terraces and in gardens.

OPPOSITE ***A magnificent tiled interior in the former convent of Santa Rosa at Puebla in Mexico. The tiles are fairly rough and ready, some with cheerful star and flower decorations, and have been used to cover every available surface, including the vaulted ceiling and the fronts of solid counters running round the room. The visual impact is made by their use over such a huge area and by their fresh colours.***

LEFT *The pattern and colour of the blue and white tiles used here to face the counter draw the eye in a room which is otherwise undecorated beyond a rough coat of whitewash. The inhabitants probably chose tiles because they were a hardwearing, inexpensive and above all practical covering.*

Popular prints from France and Italy, religious stories and themes, textile designs, even Hindu motifs drawn from imported Indian calicoes, were all represented figuratively on the Portuguese tiles, while abstract decoration took the form of energetically robust curling patterns and borders. Ironically, the tiles developed this richness and originality only when the industry became centred on smaller towns and workshops in the wake of the country's political and economic decline in the late sixteenth century. Economic recovery came at the end of the seventeenth century, and with it a mass of commissions for palaces, houses and gardens.

Brazilian gold played an important part in the dynamism of Portugal's tile industry in the eighteenth century, which benefited from the sumptuous wealth of the ruling and mercantile classes and their corresponding desire for magnificent display. Tiled rooms became increasingly flamboyant, opening up wall space visually with broad landscapes and daring use of perspective. During this period, blue and white decoration was especially popular, but renewed interest in a wider palette returned in the middle of the century. All the styles that dominated the arts in Europe in following centuries, through the baroque, rococo and neo-classical periods, were reflected in Portugal's breathtaking *azulejos*.

OPPOSITE *Modern tiles like these, hand-painted with harlequin figures against a white crackle glaze background, recapture the brilliant colours of maiolica.*

RIGHT *Splendid maiolica panel from a grand Portuguese building. The Portuguese decorated whole walls and rooms with maiolica tiles, and still do, devising magnificent designs which cover dozens, sometimes hundreds or thousands of tiles. This panel shows a river god, the precious water gushing from the amphora at his shoulder.*

In Italy, meanwhile, maiolica ware reached equally magnificent heights of achievement but in different spheres. The fine modelling and colour of the maiolica sculptures and sculptural plaques made by the della Robbia family of Florence, showing scenes in the life of the Holy Family and other biblical subjects, deserve comparison with the greatest fine art of Renaissance Italy. The della Robbias also made decorated tiles, but this was not the work for which they were best known.

Italy differed from Spain and Portugal in that tiles were primarily, though not exclusively, used on floors rather than on walls and ceilings. The brilliant effect of the most magnificent of these tiled floors, laid in wealthy churches, homes and other buildings, owes as much to the inventive combination of different-shaped tiles (reminiscent in some cases of English medieval tile mosaics) as to the lively decoration painted on to them with coloured tin glazes.

Many of the floors incorporated hexagonal lozenges, either used exclusively or in combination with other shapes. A floor in the Giustiniani Chapel in Venice groups lozenges around square tiles, with triangles in each group, to form a larger square. A particularly beautiful floor at San Giminiano has six lozenges painted with leaves and lemons on a white background, grouped around a regular hexagon of pale blue with a rosette at its centre painted in manganese on white. The tile border of this floor has cornucopias spilling fruit, verdant green leaves and yellow palms.

In Rome, the Vatican Palace has an intriguing and beautiful floor consisting entirely of diamond-shaped tiles radiating in star-like groups of six from a grid of points, and the cathedral in Siena has an even simpler but equally effective floor composed entirely of regular triangles, each with a painted rope border and yellow crescent on blue in the centre. More elaborate floors incorporated circular tiles, around which others fitted either in a regular grid of bars framing squares, or in a complicated pattern of curved pieces which must have severely taxed the ingenuity of the designer and tile makers. At Deruta, one of the Italian cities at the forefront of maiolica production along with Florence, Siena, Casteldurante, Perugia and, above all, Faenza, one special floor is covered with tiles of two interlocking shapes familiar from Islamic tile art. Eight-pointed stars are linked with crosses, each arm of the cross ending in a point. These Islamic echoes notwithstanding, the decoration is entirely Christian, with a saint or putto in the centre of each star. These and some of the floors incorporating circles (reminiscent of English medieval floors) are a reminder of the way in which patterns recur throughout the history and geography of tile-making as new designers and craft tilers try out familiar styles.

In countries across Europe in the sixteenth century, Italian maiolica tiles were hugely popular among those who could afford them, a fact not surprising considering their beauty and vitality. The enthusiasm led to most countries' setting up their own industries. In France one of the Rouen potteries was led by a craftsman who had worked with Girolamo della Robbia, and in Paris the modern formal gardens of the Tuileries mark the spot where a tile-works factory stood until 1664; now all that remains is the name. Industries were also founded in Austria, Germany and – most significantly, as the next chapter demonstrates – Antwerp in what was then the Netherlands.

ABOVE *These charming modern minstrel tiles are made in Moustier, France, in the style of pottery made there in the seventeenth century.*

CHAPTER FOUR

DUTCH and ENGLISH DELFT

MODERN HOLLAND IS ONLY A FRACTION OF THE SIZE OF THE COUNTRY THAT WAS known as the Netherlands when tin-glazed ware began to be made there. In the early sixteenth century, the Netherlands included today's Belgium and the Belgian town of Antwerp. Links between the Netherlands and southern Europe were strong, since the Netherlands was part of the Holy Roman Empire, ruled over in the sixteenth century by Spanish kings. Two types of tile were made there for use on floors: one was lead-glazed earthenware produced by modest local craftsmen; the other was maiolica, made by the aristocrats of the ceramics industry, Italian craftsmen like Guido Andries, who emigrated from Casteldurante to Antwerp in the sixteenth century and introduced their sophisticated techniques of preparing, glazing and decorating tiles.

OPPOSITE *Modern delft tiles decorated with figures of Saracens, the fierce Moorish warriors probably seen by itinerant potters who travelled north to Rotterdam in about 1610, taking the designs with them. Original delft Saracen tiles are very rare and valuable.*

At this time even lead-glazed earthenware tile floors were laid only in the homes of wealthy subjects, in public buildings and by the church. Maiolica floor tiles, which were more expensive and much less durable, were the province of only a small number of churches and nobles of great wealth, who enjoyed them for their novelty as much as for their beauty. Their appearance was identical to the maiolica discussed in the previous chapter: brightly and vigorously painted decoration, usually in blue, yellow, green and manganese, on the sparkling white ground that tin glaze provided.

Towards the end of the sixteenth century there was an explosion in demand for tin-glazed tiles. The popularity of maiolica had grown so much in previous decades that craftsmen who were originally involved in producing lead-glazed tiles switched to tin glazes, without at first changing their decorative style. Existing workshops were therefore in a position to satisfy this demand to some extent, but numerous new factories opened.

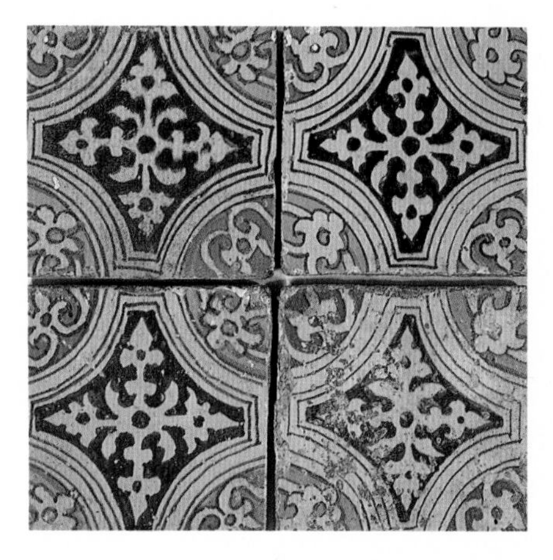

ABOVE *These Dutch tiles of around 1580 show clearly the connection between maiolica and delft, not only in the bright colours but also in the strong lines of their decoration.*

Two important reasons for the rising demand for tiles were an enormous expansion in the population, and increased prosperity, especially among a fast-growing middle class. The population increase was the result of political upheaval in response to rule from Spain, taxation without representation, and suppression of Protestantism by the Roman Catholic King. The northern part of the Netherlands was a Protestant

RIGHT *Not all Dutch tiles showed seafaring scenes or peaceful activities. These modern tiles are based on antique examples with fierce warriors in battle, capes and feathers flying, horses rearing and plunging. Others showed infantry armed with pikes and muskets. These pictures tiles are interspersed with plain tiles with only the spider's head motif decorating the corners.*

stronghold and finally became a separate state, the United Provinces of the Netherlands. In the meantime, a steady flow of people had moved from the Roman Catholic south to the safety and sympathetic environment offered by the north. Among them were many wealthy Protestant families, professionals and entrepreneurs, together with craftsmen such as sophisticated Antwerp tile makers.

The money and spirit of enterprise these people brought with them gave an extra boost to the successful and growing economy. Natives of neighbouring Germany were also attracted to the Netherlands and further swelled the population. Prosperity was capped by the establishment of the Dutch East India Company in the early years of the seventeenth century. This company imported the goods that, more than anything, influenced the style of tiles and ceramics that has made Holland perennially famous – Chinese porcelain decorated in blue and white.

Another factor in the huge growth in demand for tin-glazed tiles at the end of the sixteenth century was interior fashion. Maiolica tiles were far from ideal as a floor covering since the glaze and decoration were easily damaged underfoot. But when attached to the wall, their delicacy ceased to be a problem. In about 1580 maiolica wall tiles were introduced to the market and proved a popular and commercial success. National prosperity meant that a large number of people could afford them.

OPPOSITE *Decorated in blue, orange and green on white, these tiles are typical of the early seventeenth century. The fruit and leaves are contained within a robust quartrefoil shape, the corners decorated in relief, with the pattern of stylized foliage outlined in blue against the white background.*

ABOVE *The subjects shown on pictorial delft tiles vary hugely. Landscapes and seascapes are popular subjects, here including ducks, people fishing and boats sailing.*

OPPOSITE *An interior in the Low Countries showing tiles lining the walls. When picture tiles are used over a large area, the corner motifs combine to make a strong contribution to the overall decorative effect, which is further varied by alternating squares of four tiles with decorations of different colours – blue and manganese.*

And in practical as well as decorative terms they transformed the inside of the home. Moreover, in the western Netherlands and coastal regions, around the wealth-generating ports and trading posts (where much of the population growth in the later sixteenth and first half of the seventeenth centuries was concentrated), the ground level was more often than not below the water table and houses were consequently wringing wet, especially where walls met floors. At that time, exterior walls were without insulation or protective cavities, and indoor walls were plastered and given a coat of whitewash. This flaked, and efflorescence meant that it brushed off on anything that touched it, which in passages and narrow parts of the house meant the clothing of the inhabitants.

It soon became apparent that tin-glazed wall tiles were both waterproof and easy to maintain. They were therefore quickly adopted to overcome damp problems by acting as skirting or base boards in rooms, and covering larger areas of wall in hallways and corridors. Tiles were also used in and around the house's fireplace, of which there was usually one large one with a high mantelshelf in the room that doubled as a kitchen and living-room at the back of the ground floor. The wall immediately to each side of the fireplace might also be tiled to protect it from grease and soot.

One of the most popular designs of this period, called *kwadraattegel*, had a square drawn within the square of the tile, but up-ended to form a diamond shape. This frame was filled with a picture, of an animal or flower vase for example (the latter so popular it was produced by at least thirty tile works). The corners outside the diamond were filled with stylized leaves painted in blue and white – the forerunner of the corner decoration on delft tiles. Another popular design incorporated pomegranates, also known as orange apples, which could be interpreted as referring to the House of Orange. This noble family is today the Dutch royal family, and provided Great Britain with its late-seventeenth-century King William III, spouse of Queen Mary, last reigning monarch of the house of Stuart.

The tiles were still recognizably maiolica, painted in several colours including a lively orange-yellow. But as the seventeenth century progressed, this palette was superseded by a more restrained one – the instantly recognizable blue and white. The Dutch East India Company's supply of Chinese porcelain was not the only significant influence on this development. Fashions in clothing and interior decoration were veering towards the sober and subdued – a mood which fitted the restraint of a simple blue and white palette exactly.

Decoration on tiles became quieter: the busy diamond shape was frequently replaced by a cool circle containing a figure or figurative scene; and the decoration in the corners was reduced to an elegant, spidery motif in place of the bold and heavy leaves fashionable in former times. This trend towards the restrained can also be seen in the fine art of the time, with Rembrandt demonstrating and exploiting to the full the dramatic potential of black clothing and a near-monochrome treatment of light and colour.

RIGHT *The Dutch being a seafaring nation, it is not surprising that all aspects of the sea should feature strongly on their tiles' decoration. Here, fantastical fish cavort on the waves in a set of twelve tiles, made today but using centuries'-old techniques and the traditional spider's head corner motif.*

Besides the impact Chinese porcelain had on the decoration of tiles, it was responsible for at least three other effects on the ceramics industry. The imports included no tiles but great quantities of other ceramics. Rather than compete, the majority of factories in the Netherlands (of which there were now a great number) chose to leave aside ceramic production to concentrate on tile-making. In the town of Delft itself, two factories decided not to be defeated by the porcelain imports, but to try and beat the Chinese at their own game. They refined their materials and technique in order to produce ceramics which could compare with the fineness, smoothness and whiteness of Chinese porcelain. In the middle of the seventeenth century these technical advances were transferred to tile-making, and resulted in thinner, lighter tiles.

In 1650 a dramatic event overtook the Dutch ceramics industry. Civil war in China suddenly extinguished the trade in Chinese porcelain. The factories in Delft that had improved their technique and successfully imitated the Chinese wares stepped into the breach and were flooded with orders. Other factories in the town quickly began production, and thus began the boom that resulted in the name delft being associated with all Dutch tiles and ceramics, and even with English ceramics produced in the same style. The town did not enjoy a monopoly, however. Other centres of ceramic – particularly tile – production existed in Rotterdam, Harlingen and Makkum.

Over the next fifty years, decoration on tiles generally became sparser, though the market for more highly-decorated tiles did not disappear completely. Also during this period, more customers came from the country and fewer from the towns, where the fashion in interior decoration had once again changed. Among the middle classes, fabric and leather took over from tiles, which were relegated to cellars and the kitchen (now frequently a separate room from the main living-room).

Foreign customers, meanwhile, continued to import Dutch tiles as well as making their own imitations, and delftware held a steady place in markets across the world well into the nineteenth century and beyond. There was hardly a part of the civilized world into which delft tiles were not imported, including North and South America (especially Brazil with its long connection with Holland and Portugal) as well as European countries, of which Denmark and Germany were especially enthusiastic importers.

The range of subjects painted in a charmingly free and fresh style on delft tiles was huge, including flowers, animals, the human face (frequently of a well-known person), landscapes, seascapes, everyday life – including people's trades and occupations – and groupings of human figures. The latter often referred directly to a biblical episode, and some of these were specifically Roman Catholic, such as scenes from the life of the Virgin, and were exported to Spain and Portugal. The sea played an important part in Dutch life, and seascapes were therefore extremely popular. Ships on tiles were sometimes portraits of specific vessels; other maritime subjects such as

ABOVE *Sailing ships like these carried trading goods around the world and brought prosperity to Dutch merchants, who decorated their homes with tile panels. Some, like the lower one here, included an elaborate frame in their design.*

RIGHT *Children's games have always been a popular subject on Dutch tiles. As these tiles showing skipping, kite flying, leapfrog, swinging, and kicking a ball demonstrate, many games have barely changed. Hoop-rolling is out of fashion, but skittles were the forerunner of today's popular game of bowling, played as often by adults as children.*

sea battles, whaling and the departure of Prince William of Orange to England in 1688, also featured.

Among the most interesting subjects of decoration on Dutch tiles were children's games, which revealed much not only about the pastimes of childhood over several centuries but also about society and religion of the period. Some tiles, for example, made moral points such as warning against pretentiousness (hobby-horse riding) and about the transitory nature of temporal life (blowing bubbles). Others showed cherubs indulging in the same games enjoyed by their earthly siblings, such as rolling the hoop. Some games are no longer played today (flying live birds with a cord tied to one leg), while others like golf and kite-flying are still instantly recognizable. Interestingly, while tiles showing children (and indeed adults), playing games were hugely popular in the Netherlands and abroad in countries which imported Dutch tiles, they were rarely made by the tile industries that were established abroad as a result of the popularity of delftware.

In Britain, delftware tiles were imported in quantity and later manufactured in several cities. So fashionable were they that Henry VIII covered floors at Hampton Court Palace with Flemish tiles. A small tile-making centre in Norwich, founded in the 1560s by Jasper Andries from Antwerp, was the forerunner of thriving eighteenth-century delftware industries in London, Bristol, Liverpool and Glasgow.

The Bristol industry was founded in the 1680s and was technically competent by the 1730s. Tile makers from London, meanwhile, moved to Liverpool and made that city a tile-making centre of importance. It is no coincidence that these cities, and Glasgow, were all ports. New products and ideas were constantly flowing through them and were readily absorbed by indigenous craftsmen. Genuine Dutch delft tiles continued to be popular, partly because they were softer than English ones. Known as 'sandy backs', Dutch tiles could be cut to fit irregular shapes and spaces, which English ones could not with such ease.

English delft tiles gradually developed a style of their own, and the products of each city had their own distinctive elements, Bristol, for example, produced tiles with a unique type of border decoration, with one shade of white glaze painted delicately over the white background glaze of slightly different tone. This is known as *bianco-sopra-bianco* (white-on-white). Liverpool tiles often had polychrome decoration, while London tiles followed more closely the decoration of blue and white Dutch originals. The English tin glaze was very slightly blue or green because it had a higher lead content, whereas the basic white Dutch glaze contained more tin and was therefore whiter, but more likely to craze. It was also often less smooth and rich, while the English glaze was soft and glossy.

The Dutch delft industry survived and continues to this day, but the English industry was short-lived in contrast. The first blow to it was struck by two men, John Sadler and Guy Green, who developed a method of transfer printing, the precursor of nineteenth-century industrial methods of mass-producing decorated tiles. Tin-glazed ceramics of all types were also challenged in the eighteenth century by new products from the potteries of men such as Josiah Spode and Josiah Wedgwood, whose names are synonymous even today with high-quality stoneware of timeless design.

ABOVE *Clockwise from top left: warehouse, windmill, heron, net loft, cottage and cockerel are some of the subjects chosen by modern tile makers to continue the delft tradition.*

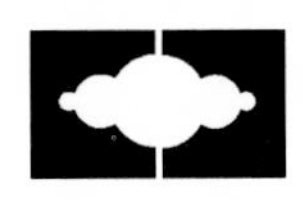

CHAPTER FIVE

VICTORIAN TILES

THE VICTORIAN AGE IS WELL KNOWN AS A PERIOD OF INVENTIVENESS AND enthusiasm. It was characterized by innovative and enthusiastic use of advances in mechanization to create and promote a huge range of products that affected ordinary people's lives. Everything from railways, criss-crossing nations and continents, to the manufacture of buttons for clothing was touched by this enthusiasm. The motive was largely commercial, but success was not possible without a positive attitude to change and progress – the attitude that new ideas were exciting. At the same time, this love affair with industrial mechanization displayed by some Victorians was balanced by another sort of enthusiasm, for philanthropy, public works such as sewers and healthy housing, and the improvement of the mind and life through contact with art and man-made (as opposed to machine-made) crafts.

Nothing escaped industrialization, least of all the tile, which was produced in vast numbers. In previous periods of history the design of tiles generally developed in one particular, distinctive direction, a result of technical advances as much as the popularity of a particular visual style. In the Victorian period, the application of various industrial methods resulted in tile-making taking off in a dozen different directions, its trails making quite as elaborate a pattern as the railways across Europe. Trying now to keep track of the methods of decoration, the range of subjects used in decoration, the technical advances and the companies and individuals who were designing and making tiles induces a sense of confusion and euphoria. It was an exciting time for tile-making.

If it is possible to begin with only one man, the person who probably initiated the new age for tiles was the architect Augustus Welby Pugin (1815-52). His interest, ironically, was in the manufacture and decoration of tiles which recalled medieval England, rather than in the pursuit of industrialization. By looking backwards, he hoped to enrich the present, and in particular the churches and grand buildings he was commissioned to design or advise upon. Pugin was gifted and energetic, and achieved success as an architect at an early age. A convert to the Roman Catholic faith, he was passionate about the Christian significance of Gothic architecture in the Middle Ages.

ABOVE *Modern encaustic tiles of the type popular in Victorian England.*

OPPOSITE *Tiles by the brilliant designer William De Morgan, showing magnificent creatures against swirling vegetation. The red lustre, inspired by Islamic tiles, is typical of De Morgan.*

ABOVE *Various designs of modern encaustic tile, made from a composition of stone and marble. As with Victorian encaustic tiles, the pattern and colours go right through the tile which makes it exceptionally hard-wearing. Modern encaustic tiles are porous and need careful sealing.*

Pugin adopted the Gothic style for the contemporary buildings he designed. His influence was enormous, his active years coinciding with the rapid increase in church-going and church building that followed the growth in urban populations caused by the industrial revolution. And he was not only concerned with the buildings; he involved himself with every detail of the furnishings of the interiors, as did later influential designers such as William Morris and Charles Rennie Mackintosh. Thus he designed not only the exterior detail of the Houses of Parliament at Westminster but also the hat stands and inkwells used by Members of Parliament and peers inside the building.

Tiled floors were the only appropriate floors, as far as Pugin was concerned, and the only appropriate tiles were of course earthenware, inlaid tiles in the medieval style. These were known in the Victorian period as 'encaustic' tiles, and Pugin's passion for them prompted a contemporary to say: 'Among the various objects occupying Pugin's attention, not one received a greater share than the revival of the manufacture of encaustic tiles.' This commitment, combined with that of the tile manufacturer Herbert Minton, could be said to have laid the foundation of the massive Victorian tile industry.

As well as the new churches that were being built, many existing churches were restored in the new enthusiasm for Gothic architecture. Considered ugly in the eighteenth century when cool classicism held sway, many medieval tiled floors had simply been boarded over and thus preserved, to be uncovered by the Victorians. Such floors, as well as those in public places like the Chapter House of Westminster Abbey, provided models for the design of new encaustic tiles. Minton's firm led the way in their production.

Two years before Victoria came to the throne in 1837, Herbert Minton had bought up the moulds, equipment and stock of an unsuccessful and disillusioned inventor, Samuel Wright. He also bought a share in the patent on Wright's invention, a mechanized method of manufacturing inlaid earthenware tiles. This method involved pressing so-called 'plastic' clay into a mould at whose bottom was a raised

LEFT *Victorian Gothic revival tiles made in 1873 by the Campbell Brick & Tile Co. and used to decorate a church in Scarborough, Yorkshire. These tiles were made by the dust-pressing method, the most frequent method of tile manufacture in the nineteenth century. It involved stamping powdery clay 'dust' into a mould under great pressure; the dust had just enough moisture to stick together, and dried quickly.*

pattern. The indentations left in the clay by the raised pattern were subsequently filled with liquid clay (called 'slip') of a different colour, resulting in a patterned tile very much in the style of medieval floor tiles.

The term 'plastic' refers to what most people think of as ordinary clay – a natural material rather like solidified mud. The term is used to make a distinction between this method of tile production and the later method which took over from it and revolutionized tile production, the dust-pressing method. Minton bought a share in the patent for this in 1840, when it was being used by its inventor, Richard Prosser, to make buttons.

By taking clay dust and pressing it into a mould under tremendous force, it was possible to make tiles that presented few of the technical problems which arose in

RIGHT *A set of four Minton tiles designed by John Moyr Smith, one of the great tile artists of the nineteenth century, made in about 1875 by the dust-pressing method, with decoration added before the final glaze. The four seasons were one of many popular themes which enabled tiles to be made in sets for use together or separately.*

the plastic clay method. The clay dust had just enough moisture in it for it to stick together when compressed in the mould, but needed very little time to dry before being ready for firing. By the end of Victoria's reign, few tiles were made by any other method. In 1844 Wright's patent came up for renewal and was bought by Minton in partnership with another tile-manufacturing company, G. Bar & Fleming St John.

Tile production became a source of fascination to the highest in the land as well as to foreign dignitaries. In the previous year, Prince Albert, the Prince Consort, witnessed a demonstration given by Minton's chief engineer at a soirée held by the Marquess of Northampton. Also present were the Duke of Wellington, the Prime Minister, Sir Robert Peel, over two dozen foreign princes and, also of great importance considering their huge potential for patronage, several bishops. Prince Albert was deeply interested, and the following year Minton was commissioned to tile the spacious corridor being built by Thomas Cubitt to connect the main block and the pavilion of Osborne House, Queen Victoria's summer family home on the Isle of Wight.

Minton's success was assured. The Minton Hollins catalogue of 1881 (by which time the company had split into several parts, of which this was the main tile manufacturer) gave a long list of prestigious commissions fulfilled. Not only does the list make entertaining reading; it also reveals the range of structures (not only buildings on land) and extent of overseas orders which Minton supplied. The list includes: 'The Royal Palaces of Windsor, Osborne, Clarence House, Sandringham and Marlborough House; the Imperial Palace of the Emperor of Germany; the Palace and State Yachts of the Sultan of Turkey; the Royal Residence of Prince Dhuleep Singh; the Houses of Parliament, Westminster; the New Foreign Offices; the New Government Buildings in India and Australia; the South Kensington Museum; the Albert Hall; the Senior and Junior Carlton Clubs; the Cathedrals of Ely, Lincoln, Lichfield, Gloucester, Westminster, Wells, Glasgow, Armagh, St Giles' Edinburgh. Dunblane and Sydney (New South Wales); The New Capitol of Washington (US of America); the Town Halls of Liverpool, Leeds, Rochdale, Bolton, etc.; the Municipal Buildings of Birmingham and Liverpool; and many of the principal Ducal Mansions, Government Buildings, Churches and Public Institutions of Great Britain, etc.' That last 'etc.' is a delightful throwaway ending which seems to say, 'And just about every other type of important building you care to mention . . .'

From the greatest building to the smallest, tiles quickly became a necessary element, more than just a decorative detail. The Victorian period, especially the 1880s and 1890s, saw a huge amount of speculative building in the suburbs of cities. Street upon street of houses for workers of every rank sprang up – terraced and semi-detached houses largely of great charm which are still coveted homes today. Hardly one was without a tiled path, porch, hallway or passage, at the very least. Tiles were used almost as extensively on walls as on floors. In small houses the spaces involved were a fraction of the areas in a palace, church or town hall, and it was the builder rather than an architect, owner or dignitary who actually chose the tiles to be laid. For these customers, the geometric tile was introduced.

Instead of the pattern being inlaid into square tiles, it was constructed *in situ* from separate tiles of geometric shape, each one a different colour, which fitted together like the pieces of a puzzle into any pattern the builder or owner cared to invent. The method is reminiscent of medieval tile 'mosaic', although the fifteenth-century monk could not of course pick his pieces from a catalogue. The choice of shapes was now huge: a page from the Campbell Tile Company catalogue of 1885 shows 109 different, compatible tiles with straight and curved edges, cut from a six-inch square. Geometric tiles could be and were used in conjunction with inlaid encaustic tiles on a single floor, as can be seen in many parish churches.

ABOVE *These tiles were all made with 'plastic' clay, in its wetter form, rather than drier clay dust. The designs show a strong interest in medieval tiles. The lion was made by Minton in about 1850 to a design by AWN Pugin for the Palace of Westminster.*

It was not only homes, palaces and government buildings which were tiled. The inside and outside of shops, offices, workshops, libraries, public lavatories, railway stations and underground stations were faced with tiles, often incorporating the name of the proprietor or place. Laws governing building standards were introduced, and advances in science and attitudes to medical care also played a part in promoting a more positive attitude to hygiene than had previously existed. Hospitals and food shops like butchers and fishmongers embraced the glazed tile as the ideal hygienic wall and floor covering. The result was magnificent interiors like Harrods' food halls in London (a late-Victorian creation, completed in 1901) designed by W. J. Neatby for the firm of Doulton. Just as medieval church floors had been covered up when out of fashion, so the Harrods food halls were hidden behind twentieth-century lowered ceilings and streamlined cladding, to be rediscovered and restored to their full glory in the late 1980s.

The most popular method of decorating Victorian tiles was by transfer printing, a technique inherited from Sadler's eighteenth-century printing system. This involved inking the decoration on to a thin piece of paper, either from an engraved metal plate or from a lithographic stone, using oil-based inks. The tissue was then laid on a once-fired biscuit tile with the ink next to the clay, rubbed, and soaked off in water leaving the design on the tile. A low-temperature firing fixed the inks before a final glaze was added and fired. Extra colour could be hand-painted over the transfer decoration either after or, more usually, before this final glaze. Transfer engravings produced decoration with precise lines; lithographs were better for designs with solid areas of colour.

Transfer printing was just one of many methods of decoration with which the manufacturers of wall tiles experimented. Painting by hand was the freest, though this was sometimes preceded by 'pouncing' the outlines of a design through tiny holes in paper with powdered charcoal. Another type of decoration, tube lining, involved squeezing thin lines of clay on to the tile, like cake icing, and filling the separate sections thus created on the surface with glazes of different colours. Relief tiles had the subject impressed into the clay so that the glaze pooled in the deeper areas, making a darker colour, while shallower areas showed a paler shade of the glaze colour. The most modern decorative technique was photographic and used mostly to transfer shallow relief likenesses of famous people on to wall tiles.

Pâte sur pâte also worked on a similar principle to relief decoration, but the thicker areas of glaze were built up gradually on the flat tile by being painted on by hand, a painstaking process. By comparison, sgraffito was a fairly primitive decorative method. Coats of contrasting slip were laid on the tile and the design scratched through the upper layer or layers, revealing the colour beneath. Aerography was a little-used method of spraying glaze through a stencil on to the tile surface, which resulted in flat areas of colour which can be seen on close examination to have a speckled quality and sometimes a slight blurring of the edges.

LEFT *The tiles for the food halls at Harrods store in London were designed to be the most magnificent example of their kind, as well as being practical and hygienic. Hidden by false ceilings and partitions in the twentieth century, they have now been restored to their former glory.*

RIGHT *Another, larger, set of dust-pressed Minton tiles designed by John Moyr Smith and made in about 1875. The theme of the set of twelve tiles is Sir Walter Scott's* Waverley *novels, hugely popular historical romances. The title of each novel appears above, the names of the characters below.*

OPPOSITE *A collection of tiles by the company WB Simpson and Sons from the later nineteenth century, including another version of the four seasons. The three single tiles show fruit, flowers and a swallow in flight. The three borders have an underwater theme, with the curving lines of seaweed, fish and shells flowing from one tile to the next.*

SPRING
SVMMER
AVTVMN
WINTER

ABOVE *A complicated arrangement of tiles, with some set straight and others diagonally and separated by narrow strips. Made by Maw & Co. for the 1867 Paris exhibition, these glazed tiles would cover the dado or lower part of the wall.*

Lists and analyses of the subjects shown on Victorian tiles fill whole books – almost nothing escaped. Flowers were immensely popular on tiles, as in every branch of the applied arts in the Victorian era. They appeared in realistic form, flattened and stylized in more schematic form, incorporated into geometric patterns, and in Islamic form in tiles with a so-called 'Persian' theme. Tiles were even transfer printed to imitate embroidery of flowers on canvas. Following on from flowers were leaves, fruit, birds, animals local and exotic, insects, fish, landscapes, the elements, the seasons and the days of the week.

A large number of designs incorporated the human figure, and many of these referred to works of literature. Characters from popular fiction and contemporary poetry were balanced by mythological and Shakespearean heroes and heroines, and scenes from plays, fables and tales. Religious subjects were popular, as were improving mottoes such as 'Waste Not Want Not', which was probably incorporated into the kitchen tiling as admonishment to the servant or housewife. For children there were cherubs, fairies, songs and nursery rhymes, with some of the period's most celebrated illustrators, such as Kate Greenaway and Walter Crane, contributing designs. Many accomplished artists, whose names are not as famous as these, worked for the tile manufacturers, and as a result much tile decoration was lively and charming.

Painted panels, the largest incorporating dozens of tiles, decorated shops with scenes considered appropriate to the goods or food being sold – cows or a bull in a landscape for a butcher's, for example. And hospitals, especially children's wards, were similarly decorated but with scenes calculated to cheer the patient, such as nursery rhymes and lush pastorals. Sporting scenes on panels and individual tiles varied, from the heat of the chase to the gentle game of croquet played on smooth lawns, and even chess. Figures in these were as likely to be in medieval as modern dress. The trades were a popular subject, as was the idyllic portrayal of activities connected with agriculture. Food and meals decorated many tiles, as did associated mottoes and homilies such as 'Eat thy food with a thankful heart', and 'A little pudding adds to the repast'.

Machine-based industrial production of tiles made this huge choice available to millions of people. William Morris and his followers, on the other hand, proposed that the machine could not be used to produce any object of true artistic merit. The irony that this position made art elitist by putting it beyond the means of most ordinary people was not lost on him and he later softened this view to allow the machine to be a tool in the artist's hand. Today he is famous primarily as a designer of timeless wallpapers which are still in production. In his own age he was a prominent man of ideas. His views provided leadership for many artists and craftsmen who found mechanization unsympathetic or incompatible with their work.

The movement he lead away from machine production was known as 'Arts and Crafts'. In another interesting twist in the history of tiles, Arts and Crafts designs and ideas in general were influential in Holland. Dutch methods of production were small-

LEFT *This exuberant panel is by William De Morgan and shows his great interest in Islamic designs. The interlacing plant forms and vivid shades of blue on white are reminiscent of the swirling decoration on Iznik tiles, while the chrysanthemum borders are instantly familiar as De Morgan.*

scale and craftsman based, rather than large-scale industrial, so the Dutch tile industry was fertile ground in which Morris's ideas thrived. Many Dutch tiles of the period could be mistaken at first glance for Morris creations.

Tiles which British Arts and Crafts artists decorated were not wholly successful, either technically or decoratively, and it was left to one particular artist, William De Morgan, to carry the banner into the world of ceramics. Trained as a painter, De Morgan came to tile-making via involvement with Morris and experiments with stained glass. He was a single-minded genius whose designs are among the finest artistic creations of the Victorian age but whose business acumen was almost nil. He ended his life as a successful novelist, still producing magnificent designs but forced by ill health to live abroad, divorced from his (now defunct) ceramic works in London.

ABOVE *Modern tiles made by Kenneth Clark Ceramics to designs by the Victorian artist William De Morgan, as a result of a revival of interest in the period. Camilla Clark has also designed tiles in De Morgan's style.*

William De Morgan produced expensive tiles with wonderfully dynamic and colourful designs. His most frequent subjects were plants and flowers, exotic and mythical birds and beasts, and ships with billowing sails, often boldly decorated, skimming across choppy seas peopled with porpoises. The forms are bursting with energy, almost too much so to be contained by the edges of the tile. Any spare space around the creatures is almost invariably filled with semi-abstract tufts of grass, sprigs of vegetation or swirling leaves, always exquisitely drawn and taking nothing away from the central figures.

De Morgan, like other Arts and Crafts artists, was strongly influenced by the designs and colours of Islamic tiles. His favourite palettes included either strong blues and greens familiar from Iznik work, or a distinctive rich red lustre. The colours stand out brilliantly against the strong white glaze which De Morgan applied to blank biscuit tiles made from plastic rather than dust clay.

Although Arts and Crafts designers – apart from De Morgan – may not have been adept at making or decorating tiles, they used tiles enthusiastically in an applied context. In particular they incorporated them into pieces of furniture, sometimes ordering them from Holland or Turkey for the purpose. Arts and Crafts architects like Philip Webb and Richard Norman Shaw faced buildings with terracotta tiles in the age-old vernacular method of tile-hanging, which they revived and juxtaposed with brickwork, stone, half-timbering and plaster.

An offshoot of the British tile industry and the Victorian era's mania for huge international exhibitions of industrial and 'artistic' products, was the American tile industry. At first, British companies dominated the American scene, employing agents to promote and sell their tiles. Companies such as Minton Hollins staged splendid displays at shows like the World of Science, Art and Industry in New York in 1853. The American authorities encouraged British craftsmen and technical experts to emigrate to a more prosperous life and bring their expertise to the youthful industries of the New World. As a result, in the last three decades of the nineteenth century companies producing tiles indigenous to the United States sprang up across the breadth of the country, and as often as not failed within years. This rapid turnover in companies means that a history of American tiles threatens to read like little more than a list of names. In fact the industry had a distinctive character of its own which distinguishes it from its British parent.

Tiles were never as common a feature of Victorian architecture in America as they were in Europe, added to which the transfer printing technique was never as popular as in Britain. The American tile industry was more preoccupied with hand-made and hand-finished 'artistic' tiles than mass-produced printed ones. Edwin Atlee Barber, author and commentator on the ceramics industry of the United States, observed that 'it is to the broad extension of the Arts and Crafts Movement that we should look for the best results in artistic development in this country'. Tile-making attracted the enthusiastic attention of artists and craftsmen as well as industrialists, and one result was the Tile Club. In New York, on Wednesday evenings over a period of ten years from 1877, a group of artists entertained themselves by decorating tiles in the studio of one or another member. Perhaps the most famous members were the painter Winslow Homer and the architect Stanford White. It seems that as much time was spent on cheerful conversation and exchange of anecdotes as on artistic endeavour, but the growing membership of the club and its decade of existence bear witness to the general interest in tiles in America. The industry survived until the depression of the 1930s dealt it a final blow.

ABOVE *Victorian-style tiles made at the historic Jackfield tile works in Ironbridge Gorge, which opened in 1883 as the largest decorative tile works in the world.*

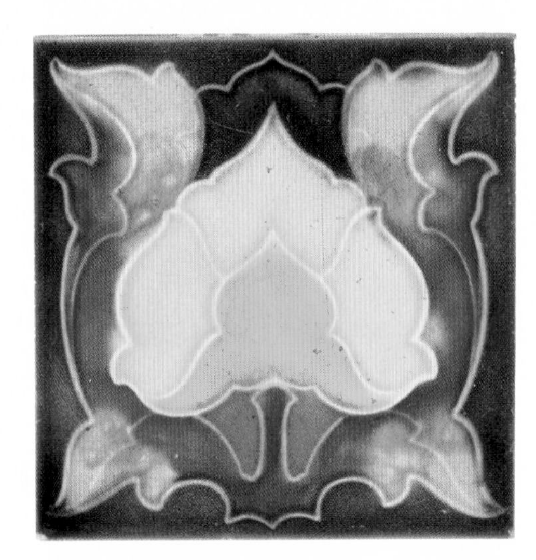
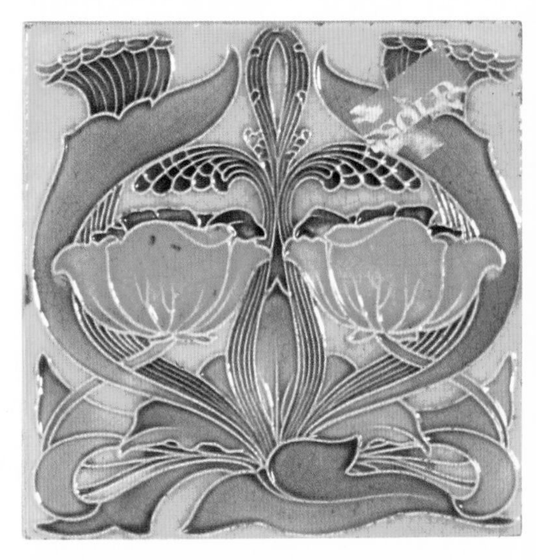

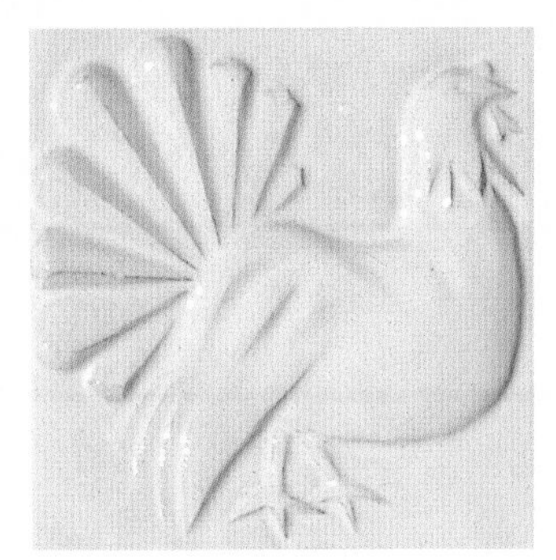

CHAPTER SIX

The TWENTIETH CENTURY

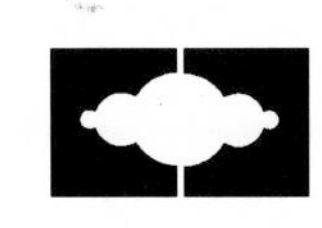

WHEN THE TWENTIETH CENTURY DAWNED, VICTORIA WAS STILL ON THE THRONE of England. Interiors looked Victorian, and had done so for decades. A hundred years later, not only is the look of people's homes radically different, but the rate of change in interior fashion is much quicker. This new state of affairs was launched in Britain by the Festival of Britain in 1951, which heralded a post-war, colourful, ration-free era. Later it was promoted by the arrival of DIY as a leisure pastime and an industry. Other factors were an increasing number of full-colour home interest magazines which found a ready market, and a preoccupation in all the media with the concept of 'lifestyle', of which the home is an important part.

Three main themes or categories which divided the production and appearance of tiles at the beginning of the twentieth century still exist, perhaps even more strongly, almost a hundred years later. These are: industrial mass-production; semi-mechanized or smaller-scale industrial, hand-finished or craft production; and individual tiles and tile panels, hand-made or hand-decorated by skilled artists and craftspeople working alone or in very small groups.

As at the end of the nineteenth century, the great majority of tiles made in Britain at the end of the twentieth century are the result of industrial mass production and fall into the first category. Victorian tiles produced in this manner, however, have a strength of character, a variety, and a charm that most of today's bland, timid designs lack. Some Victorian industrially-produced tiles were hand-coloured, which gave them an element of individuality. Today, richness of design and texture, energy and innovation are to be found not in tiles offered by the big industrial companies, but in the creations of an increasing number of individual craftspeople, who fall into the third category of producers.

These artist-designers are supported by a handful of small manufacturers and, since the early 1980s, by a few committed retailers. The latter also import craft tiles which are mechanically produced in smaller or larger quantities and are hand-finished or decorated. These tiles have usually been designed by artists, who may also be involved in the production process, and who are often continuing a local tradition of

OPPOSITE ***The art nouveau style favoured plant and flower forms which allowed the artists to draw seductive curves and elegantly interlocking forms. Most of these tiles have machine-pressed relief decoration.***

RIGHT *Six fantastic creatures by Bronwyn Williams-Ellis, including heraldic and mythological beasts, printed on unfired blanks with cut-out sponges using a method similar to children's potato prints. Sponge gives the interesting mottled effect on the animals' bodies.*

tile production and decoration which has existed for many decades or even centuries. Colourful hand-painted tiles from Spain and Mexico are examples of these, as are blue and white tiles made in the delft style in Britain and elsewhere. Hand-finished and hand-decorated tiles have a special quality, often a slight roughness of finish and always a sense of individuality, which the smooth, seamlessly repetitive mass-produced tiles do not, and they fall into the second category. A century or more ago, this would have included work by William De Morgan and the Arts and Crafts artists following the ideas of William Morris.

This transfer of creativity and interest away from the industrial, towards the individual, can be charted through the events and styles of each decade of the twentieth century. The first of the century's styles, whose sinuous lines and curving flower and plant forms flow through from the Victorian period, is art nouveau. Though industrially produced, some of the tiles decorated in the art nouveau style have a freshness and comparative simplicity, in their use both of line and of clear colour, which look forward to the new century and away from the crowded detail and moody colours of Victorian style.

By the time art nouveau was the prevalent style, however, the heyday of tile production was over and declining demand caused the closure of many tile factories. In 1914, the onset of the First World War in Europe dealt a near-fatal blow to an industry which was in decline after enlivening domestic, civic and commercial architecture and interiors for generations of Europeans and Americans.

Between the First and Second World Wars, the prevalent styles in architecture and interiors were modernism and art deco, both of which demanded exactly the clean lines and hard surfaces which tiles provided. Modernism demanded that buildings and rooms be uncompromisingly of-the-moment, uncluttered by frills and ornamentation, with straight or archly curving lines and smooth surfaces. Tiles were decorated with geometric motifs or were plain, in colours like white, black, soft emerald green or pinky-beige, and were arranged in flat bands of colour or in geometric patterns. An exception was the work of the Omega Workshop artists, amongst them Vanessa Bell and Duncan Grant. These artists, based in Bloomsbury in London, produced flamboyant, hand-painted, colourful work, including tiles, under the eye of the man who introduced London to the vibrant colours of Post-Impressionism, Roger Fry.

The inter-war period was not long enough or robust enough, however, commercially or in terms of design, to foster a new tile industry, and war intervened once again to suppress all but the essential industries. A tile industry emerged in Britain again after the war, mass-producing tiles for use in domestic bathrooms and kitchens, but it tended to use internal designers, who seemed barely to deserve the name. Kenneth Clark is a veteran tile artist who was a student at the Slade art school in London immediately after the war. He observed the situation in industry and found it depressing.

'The post-war period saw a dearth of new tile designs in industry. If what they were doing before the war could be sold then there was no incentive to change. A design policy as such was not a vital element of factory life. Staff designers if they existed were just another category of operative and were treated accordingly. The designer was often the head decorator who had some drawing skill and was able to

ABOVE *Modern tile of bluebells made at the Jackfield tile works from a metal dye which has been cast from an art noveau tile.*

LEFT *Two tiles showing women's heads drawn with the languid lines of the art nouveau style and overglazed in vivid green. The left-hand figure is based on an illustration which appeared in the magazine* The Studio *in 1901, so the tiles date from that year or later.*

ABOVE *Small square modern marble tile which has been 'distressed' by being tumbled in a drum for several hours. The delicate figure of a deer-like creature has been hand-etched with a roller-cutter and a light stain added afterwards to highlight the etching.*

OPPOSITE *Dancer, by the gifted and prolific contemporary tile artist Bronwyn Williams-Ellis. She uses a variety of techniques for decorating her tiles including this wax-resist method of drawing with a free, energetic line.*

adapt ideas suggested or supplied by the managing director or the sales representative. He was seldom trained as a designer or expected to be an innovator, but he was expected to churn out a continuous selection of designs until one caught the fancy of the person responsible for selecting what went into production. When in the late '50s and '60s design as a concept began to emerge, the designer was given more freedom – but sadly he could not come up with the goods having been so long creatively frustrated and imaginatively undeveloped. The results were rather pathetic.'

This situation described by Kenneth Clark contrasts strongly with industry's enthusiasm for artist-designers a hundred years earlier. Leading artists and illustrators like Kate Greenaway, Randolph Caldecott and Walter Crane were then employed to design tiles. Even in the first years of the twentieth century, tile designs by the brilliant architect and artist C. F. A. Voysey were in production. This is not to say that modern artists do not make tiles. Throughout the twentieth century, artists like Pablo Picasso, Salvador Dali and Joan Miró have occasionally worked on tiles, but they have been doing so for themselves or clients, not generally for industry.

The two great twentieth-century masters whose tile work is perhaps most original are Henri Matisse and Antonio Gaudí. In his old age, Matisse created a tile mural for a chapel in Vence, near where he lived in the South of France, using simple line drawings in black on gleaming white. The result, which shows the Virgin and Child and other biblical scenes, is thrillingly bold and direct. The gleaming hardness of the tiles and uncompromising black lines contrast dramatically with the strength of emotion both in the Virgin's protective figure and gentle embrace of the Christ child, and in the child's outstretched arms reaching to embrace us, as we look at him, and the whole world beyond us. Gaudí's work makes its impact in a completely different way. He used coloured tiles on the surfaces of buildings and in other architectural ways, to create a mosaic of rich tone and texture which is dazzling in its complexity. The Güell Park and other buildings in Barcelona in Spain represent some of his most glorious achievements using tiles.

One of the century's greatest potters, Bernard Leach, also made tiles occasionally. As with his pots, these show the influence of his childhood experiences of Japan and his lifelong interest in its art as well as the pottery of China and Korea. Shapes were startlingly simple and unadorned; decoration was graphic and hand-painted or incised, showing simple motifs of fish, trees or flowers, figures and landscapes. Colours were muted, with the colour of the clay showing through subtle translucent glazes or complementing soft shades of green or blue. The tiles follow this pattern. Leach had many pupils, and his influence in the art schools was tremendous. Work in the Leach style and in terracotta was nick-named 'ginger pottery' by younger artists who found it drab and who reacted against it by using bright and strong colours. Followers of Leach, meanwhile, thought the use of colour in such work garish and vulgar.

The current renaissance in tile design and production dates from the late 1970s. In the meantime, any colourful style and originality in tiles came largely from artists

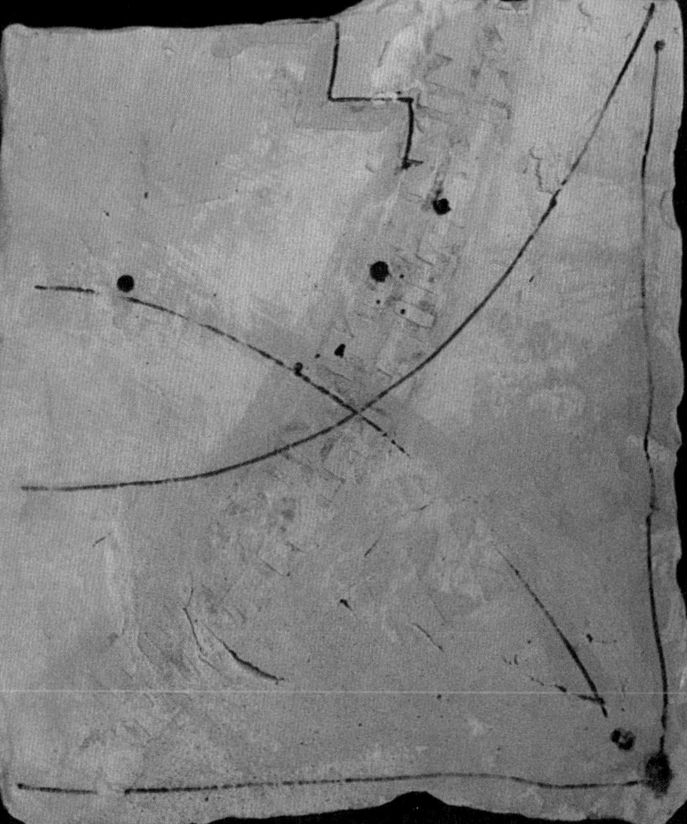

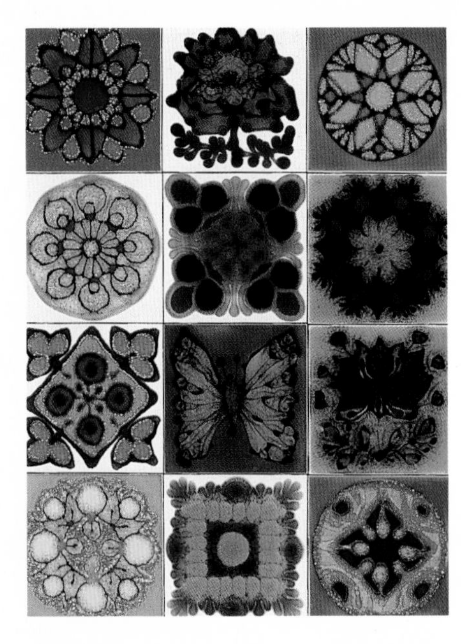

ABOVE *Twelve single tiles with colourful geometric and semi-abstract designs by Ann Clark, an established British tile artist whose work is made by Kenneth Clark Ceramics, the company she founded with her husband.*

who had trained not as potters but as textile or graphic designers or as painters, like Kenneth Clark. He was primarily interested in colour and dabbled in pottery one day a week while studying painting. But without any technical training in using glazes he had difficulty in making his colours stay on the vertical surfaces of pots. Instead, he turned to the horizontal tile, off which the glaze did not run so readily. In the 1960s the family firm, which he and his wife Ann founded, produced tiles with decoration in rich turquoise and burnt orange. They also found that customers wanted to use black tiles, mostly with red, turquoise or plain white.

Other rich colours like chocolate brown, dark green and purple were also popular. Many of these colours continued to be fashionable well into the 1970s, when more muted colours came into vogue. As well as individual plain and motif tiles, the company produced border tiles decorated with fish, leaves and flowers, and tile panels decorated with the tube-line technique popular with the Victorians. The Victorian revival in the 1980s also prompted them to revive designs by William De Morgan, and to create their own designs in his style.

Enthusiastic interest in tiles, among British art historians and amateur collectors as much as among artists working on tiles, resulted, in 1981, in the founding of the Tiles and Architectural Ceramics Society. As well as being a forum for the exchange of ideas and research, this group has become involved in conservation issues, helping protect historic tiles. An ever-increasing number of artists, meanwhile, produced tiles and tile panels in their own style, using a wide variety of techniques, derived from their own visual influences, so that the last decades of the nineteenth century show

RIGHT *A collection of single tiles by Bronwyn Williams-Ellis, stamped with relief decoration and glazed freehand. As well as single tiles, she makes panels of tiles like Dancer on page 65, and tile schemes for parts of rooms and whole rooms by special commission.*

LEFT *Panel of thirty tiles designed and decorated by Ann Clark using the tube lining method popular with the Victorians. The artist draws with a thin line of clay squeezed from a tube like cake icing, which subsequently keeps coloured glazes separate from each other.*

no one overall theme. Much of the work is bold, strong and colourful, produced for one-off selling exhibitions in art and craft galleries and for individual clients. Specialist tile shops also stock tiles by contemporary artists, though the financial rigours of business and the discipline and pressure of producing the required number of tiles of suitable quality mean that not all artist-tilers can cope with, or want to be involved with, the retail market.

Retailers have found that all types of fine tiles, including hand-finished terracotta, rough slate and smooth marble, and hand-decorated tiles of greater or lesser sophistication from all over the world, are immensely popular. The 1980s saw a boom in trade in these tiles, as homeowners were prosperous and channelled funds into beautifying their homes. Stagnation of the property market and the economies of western countries has inevitably affected trade in tiles, but among artists there is no shortage of enthusiasm for their tilework and there are still plenty of commissions and customers. No recession can alter the basic attractions of tiles in the home, both in practical and in aesthetic terms. The late twentieth century is as exciting a period in the history of tiles as any peak of achievement in previous centuries.

CHAPTER SEVEN

Floors I: FUNCTION and STYLE

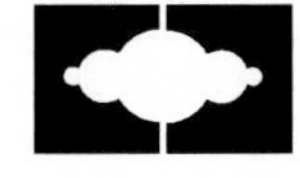

THE USE OF TILES AS FLOOR COVERING HAS A LONG HISTORY, AS PREVIOUS CHAPTERS have shown. In the western world during this century tiled floors have never completely gone out of use, but they have suffered periods of relative unpopularity. One such period, in which man-made synthetic floor coverings have been to the fore, is undoubtedly coming to an end. Though it is hardly a new concept, the idea that natural materials are generally the most beautiful and the most pleasant to live with is now prevalent in all areas of home life, from clothing to wall paints and floor coverings.

Tiled floors fulfil this idea perfectly, at the same time offering a greater choice in terms of material, colour and finish than was available in earlier centuries. Modern techniques make a wider range of tiles available for use on floors – for example, glazes and methods of firing which make ceramic tiles tougher than in previous centuries when they were considered suitable for wall decoration alone. Age-old methods of sealing and finishing tiled floors, such as sealing with boiled linseed oil and polishing with wax, have been joined by new approaches which can combine the best of old and new. Sealants and waxes which include synthetic components can provide labour-saving finishes on tiles laid in areas of the home that suffer especially hard wear – such as hallways, kitchens and children's playrooms.

Modern methods of transporting goods worldwide also mean that the home-owner in Europe or America can equally easily choose brightly-coloured ceramic tiles made in their home country, or antique French terracotta tiles, or slate tiles from Africa. Many of the techniques used in making tiles have hardly changed in centuries and, together with the tiles themselves, have stood the test of time. Terracotta tiles not only do not deteriorate, they positively improve with use; as a patina builds up on the surface they become even more resilient to wear and tear, and their colour becomes richer and more mellow.

The combined result of all these elements means simply that there is a tiled floor to suit every purpose and style of decoration. The only limiting factors are practical – some particularly heavy tiles are not well suited for use on upstairs floors, for instance,

OPPOSITE ***A floor of square terracotta tiles with glazed decorated tiles inserted at intervals. In general, the surface of glazed tiles is more delicate than unglazed, and better used on walls or, as here, used sparingly to add interest and light to a floor. This trick is especially useful in darker rooms and passages.***

ABOVE *Tiles available today include some, like these made from Indian slate, which are cut from the earth rather than being fired. Indian slate has a rugged texture and wonderful variety of colour.*

OPPOSITE *Small marble tiles with a charmingly uneven, distressed finish look as if they have been in situ for generations. The terracotta tiles beyond are antique and reclaimed for modern use. The two floors sit happily side by side because the style of each is similar; the two types of tile mixed together could look stunning.*

and glazed tiles are inevitably more vulnerable to wear than unglazed ones – but practical considerations are also those which often make a tiled floor the best choice, and have done so since Roman times and before.

If you are in any doubt about whether to lay a tiled floor, the best approach is to consider the alternatives. Of all contemporary floor coverings, that which is blindly adopted more often than any other is carpet. Wall-to-wall carpeting certainly has attractions and an image of luxury, but it also has many disadvantages. In kitchens (and many people would say in bathrooms and lavatories too) it is generally considered unhygienic. It wears out, quite quickly in the case of poor quality types, and generally becomes increasingly tawdry in appearance as it wears. It is susceptible to marking, which makes it a nerve-wracking commodity in any household with children or animals and totally unsuitable in areas where muddy boots and shoes are likely to tramp on it regularly. Moreover, it harbours dust and a variety of minute living organisms, some of which can cause distress to allergy and asthma sufferers.

Possible alternatives to carpet are linoleum, vinyl flooring and bare wooden floors. Linoleum is composed of natural materials including linseed oil and cork, which has many attractions and practical uses. But vinyl, like some other more recent man-made materials used in buildings, has come under suspicion of being less than healthy to man as it ages. There are reports that certain types may give off gases over a long period of time. In purely aesthetic terms, vinyl is unpopular with people who prefer a real wooden or tiled floor to a synthetic floor covering which imitates those materials.

The most promising alternative is perhaps bare boards, which if sealed, treated or waxed, become less laborious to look after and can take on a colour or finish. But wooden floors have their problems too. Draughts and dust can come up between the planks, small items can drop down the gaps, old boards are often crudely nailed down or cracked and broken and uneven. Noise can also be a problem with wooden floors, as they are usually suspended over a floor cavity which causes them to act like a sounding board. Tiles are certainly not the quietest floor covering, but they are not necessarily as noisy as their reputation suggests, since they are generally bedded on a solid floor which dampens resonance.

Tiles also have a reputation for being cold and hard – certainly carpet would beat them on these scores, and would be more likely to break the fall of a fragile object dropped on it. But in countries like Holland, Germany, the Scandinavian nations and the USA, where underfloor heating is not as unusual as it is in other parts of the western world including Britain, tiles are considered a warm flooring. Laid on a solid base in which the underfloor heating pipes are buried, tiles are in direct contact with the source of heat and display what is called low thermal inertia. The porosity of the earthenware tile in particular means that it warms up quickly and retains heat.

People who have lived with extensive tiled floors, even without underfloor heating, leap to their defence when tiles are accused of being cold and hard. They declare that they are no colder than other hard floors, don't admit upward draughts like

some old wooden floors, and in the case of materials like terracotta are positively warm and welcoming underfoot. In addition, they maintain that the aesthetic satisfaction that a fine tiled floor gives, and the practical advantages, far outweigh any supposed disadvantages. Tiles are water resistant, hygienic, practical, easy to clean, look good, are long-lasting and positively improve with age.

Tiles can also be used in combination with other types of flooring. In a kitchen with a wooden floor and a solid fuel cooker like an Aga, an area of floor in front of the cooker can be tiled in a colour complementary to the colour of the enamel finish on the cooker – yellow against a pale blue Aga, perhaps, and cherry red or deep blue against a cooker that is racing green. In a sitting-room or dining-room with an open fire, you can tile the hearth in plain tiles and define its edge with a patterned border, or vice versa. If there is a stone or concrete hearth beneath the tiles, this will give them a solid base and they will simply need to be bedded on adhesive and grouted. If the floor beneath is wooden, lay the tiles on an adhesive compound designed to be elastic, to allow for any movement in the sprung floor beneath.

Deciding to have a tiled floor is just the beginning of the decision-making process. The choice of tiles available is enormous and can be baffling. But as with any type of furnishing or decorating, the decision-making can be helped by going back to basics and considering, in the case of tiles, the factors on the following checklist:

ABOVE *Slips made of some other colour or material can be inserted between plain tiles to give greater definition to the geometric pattern of a tiled floor.*

- Material
- Finish
- Colour
- Size and Shape
- Pattern

Tiles are made not only of fired clay but from various natural materials which are cut rather than fired to form tiles. These include stone, slate and marble. A clay tile can be earthenware or stoneware, which refers to the highest temperature at which it has been fired. Starting at about 1,100°F, porous earthenware undergoes a chemical change and fuses or vitrifies into stoneware. Stoneware is usually glazed, which guarantees that it has a hard, non-porous finish; earthenware is usually unglazed.

Earthenware tiles are generally known as terracotta, but this term can be confusing because the word also means a warm colour which could be described as brownish-pink or earthy red. Terracotta is a word of Latin origin which literally means 'cooked earth' or 'fired earth' and does not refer specifically to colour. A terracotta tile can be terracotta coloured, because the clay is that colour before and after firing (called 'red-firing' or 'red-burning' clay), but it can also be a paler, more yellow shade. This may be because it has bleached in the sun over generations, or because it has been purposely bleached to make it paler, or because it is made from paler clay such as a type found in northern France. Today, a pale yellowy colour may also be due to chemical additives in the clay or some other form of manipulation such as blending.

Terracotta tiles today come from a number of countries, not only Italy. Spain, France and Mexico all make terracotta tiles for local consumption and export.

OPPOSITE *This elegant and dramatic floor uses three colours in a traditional chequerboard, the two darker shades alternating and toning with the fire surround. In the background, coloured glazed tiles create a vibrant, jazzy surround for a free-standing cast-iron stove. The pattern includes shapes and sizes which could be cut from whole tiles.*

RIGHT *An interesting piece of cut-work, where a pattern has been custom-made for the floor to match the cool, classical mood of the room's furnishings. Small pieces of white tile, glazed or marble, can be inserted, as here, to highlight the design. A pattern like this needs to be carefully worked out on paper before being put into action.*

Methods of production have changed little over the centuries. A certain amount of mechanization may have been introduced to the early stages, or to all stages, of production, and the supplier from whom you buy terracotta tiles should make it clear whether or not your tiles are hand-made. Reclaimed terracotta has generally been made by hand because of its age, but check this too if you want to be sure you are buying a craft, rather than a mechanized product.

There is often confusion about the difference between terracotta and quarry tiles. In fact the two are made from different clays, but the most significant difference is that they are fired at different temperatures. Quarry tiles are fired to a much higher temperature, which causes them to vitrify to a certain extent. This means that they are less porous than terracotta, much harder, less prone to size variations between

LEFT *Highly polished terracotta or quarry tiles provide a practical and hygienic floor in any kitchen, especially in a household with pets, though they are harder underfoot and under dropped china than other floorings. In a room with massive proportions such as this magnificent kitchen, be careful to use tiles that are in scale with the surroundings.*

ABOVE *The pattern of this floor consists of sophisticated octagonals with small square inserts, in contrast to the material which is rugged African slate. The result would be equally at home in town or country.*

OPPOSITE *Plain white low-sheen glazed floor tiles make an impressive contribution to this stunning all-white city interior. Difficult to keep spotlessly clean, white is a colour that does not mix well with dogs, children and Wellington boots.*

one tile and the next, and less likely to be warped. They look, and indeed usually are, machine-made, which is not to say that they are not as good quality as terracotta, simply that hand-made and hand-finished tiles look different and more rustic. The appearance of quarry tiles is flatter than terracotta, less rustic and more regular. They may also be glazed.

The finish and texture available today on tiles varies from the high gloss of a thick smooth glaze, through the metallic sheen of a lustre glaze, to a matt, unglazed surface which may be rough or smooth. Glazes vary in strength (darker colours tend to be softer), and some glazed tiles are generally less hardwearing over many years than unglazed ones. Ask your supplier for detailed information in this respect. Unglazed earthenware tiles are matt in their natural state and need to be sealed if they are to be resistant to dirt and grease. Some are sold pre-sealed – a detail which is worth checking with your supplier. If they are not already sealed you can do this yourself without much trouble, or get a professional to do it. Alternatively you can simply leave them to absorb the dirt for a rapid aged, distressed effect (see chapter twelve, Practicalities, for basic information about sealing). Some stone and slate tiles may also need sealing if you want them to retain something like their original, unmarked appearance. They may also have a rough rather than a smooth matt surface.

Tiles are available in a mouth-watering array of colours. Apart from the relatively few types of glazed tile that are suitable for use on floors, the colours you can choose from tend to be the colours of the natural materials from which the tiles are made. There are exceptions to this – encaustic tiles for example, where the colour is not applied to the surface but goes right through the tile. If you do decide to limit yourself to unglazed tiles, the range of colours is still tantalizing, from green, blue and grey slate, through the many colours of marble, to the rich variations of reds, pinks and yellows in terracotta.

Just as you would when choosing paint for the walls, use the colours of a tiled floor to blend or contrast with other colours in a room, such as curtains, chair and sofa covers, pictures and objects. In a kitchen or dining-room, do not overlook the colour or colours of the china off which you eat, as this is an important part of the decorative scheme. If the tiled floor is to provide a background for rugs, take their colours and the style and intricacy of their patterns, if any, into account too. Look out of the windows of a room and consider whether you want the colour of your floor to contrast or blend with what you see – be it a brown roofscape, massed buildings in red brick or grey stone, or a lush landscape of green fields.

White is of course a useful and popular colour in interior decoration, but needs to be used carefully on floors because it shows dirt and marks more dramatically than any other colour. The drama of white means it can be used to stunning effect, and will be less trouble in practical terms in a household without children, animals, or direct access to the outdoors, such as a city flat with an internal entrance. In a house,

ABOVE *Rectangular terracotta tiles enlivened by a small, colourful, glazed, hand-painted insert, in character with the hand-painted cupboard alongside.*

RIGHT *A floor of polished slate like this, if carefully laid, will give care-free service and last generations, looking as good decades from now as it does today. The dark colour looks serious and handsome and won't show dust in much-used halls and passages.*

especially in the country, it is probably best suited to an upstairs bathroom, a room in which people will usually be wearing soft shoes or slippers.

All these elements in a tile's appearance – material, finish, texture and colour – together with others like shape, size and pattern (which are discussed in the next chapter, Floors II: Design Directory), must of course be related to the function and style of the room in which the tiles will be laid. This in turn relates not only to the overall decorative effect of the room but also to the style of your whole home, and its period. In the country, is your home a rough-and-tumble country cottage, or does it have a sophisticated interior with fine antique furniture? Perhaps it is a combination of the two, with some family rooms and others containing your more precious and fragile furniture and possessions. Does it have low beamed ceilings, or is it a Georgian rectory or townhouse with high airy ceilings and tall windows which flood the rooms with light?

In a town or city, is your home a Victorian terrace house, or a cosy attic flat? Is it a twentieth-century period home – art deco perhaps – or a high-tech modern space for living? Does it combine living and working areas with different practical and aesthetic requirements? Take a metaphorical step back from your home and consider it objectively in terms of its form and function. You will probably find that decisions about the type of tile to lay in various rooms and spaces become easy to make, or make themselves with little or no effort on your part, leaving you to enjoy the pleasure of exploring the great choice of tiles available.

ABOVE *Reclaimed terracotta is available in many colours and shapes. These French tiles, top, are a darker, richer red than the softer, browner shade of the Italian terracotta rectangles above.*

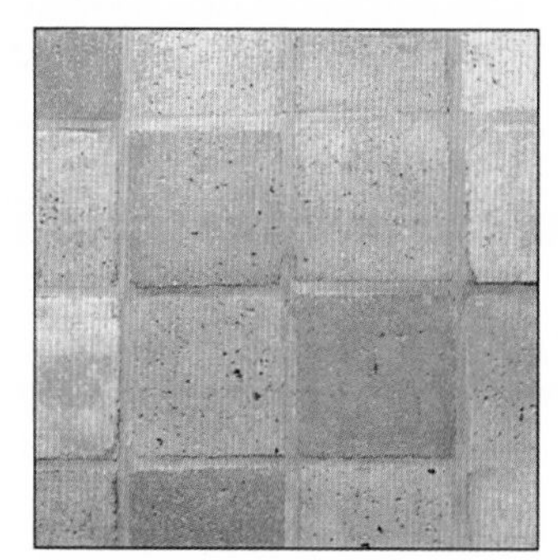

CHAPTER EIGHT

Floors II: DESIGN DIRECTORY

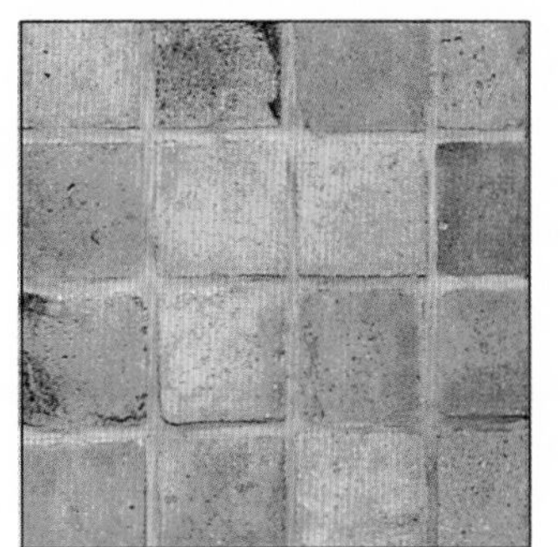

WHEN YOU WALK INTO A ROOM, THE STRUCTURAL ELEMENT MOST EXPOSED IS usually the floor. Walls have furniture standing against them, with pictures or cupboards above, and are interrupted by doors and windows. One's eye is rarely led directly up to the ceiling, unless it is decorated in a special way. The floor, however, spreads out before you. Its visual importance makes it a powerful element: it can make a positive contribution to the success of a room's decoration, or it can make a mess of it. Alternatively, it can retire quietly and discreetly, leaving some other element of the room to make an overriding visual statement.

Pattern is clearly significant in this context, both on the individual tile and over the floor as a whole. Use it carefully to avoid fussiness, especially if you plan quite an elaborate floor. Conversely, use pattern to give interest to a plain and simple or empty room. In either case, consider whether these are the only tiles in the room: in a kitchen or bathroom, for example, there may well be other tiles, on walls or other surfaces. If there are to be other tiles, plan the tiling as a whole, linking the different areas to each other and possibly limiting pattern to either the floor or the walls.

A few types of floor tile have pattern on each individual tile – encaustic tiles, for example, and inlaid, medieval-style, craftsman-made tiles. Subtle pattern on a floor may also be the result of the natural colour variations and texture of the material from which the tiles are made. Organized pattern over the whole floor is usually the result of careful planning. For this you should take plenty of time and equip yourself either with suppliers' catalogues and a book such as this one, which clearly illustrate the wide choice of possible patterns, or with plenty of paper, pencils and coloured pens, if you want to design your own floor from scratch. Suppliers' samples, which will give you an accurate idea of size, colour and finish, will also help you design your floor.

An initial consideration is the size and shape of tiles you wish to use. There are numerous sizes and shapes of tile available. The most frequent and versatile is the plain square in various sizes. Square tiles are usually laid together in a grid, with sides either parallel to the walls of the room or diagonally, at forty-five degrees to them

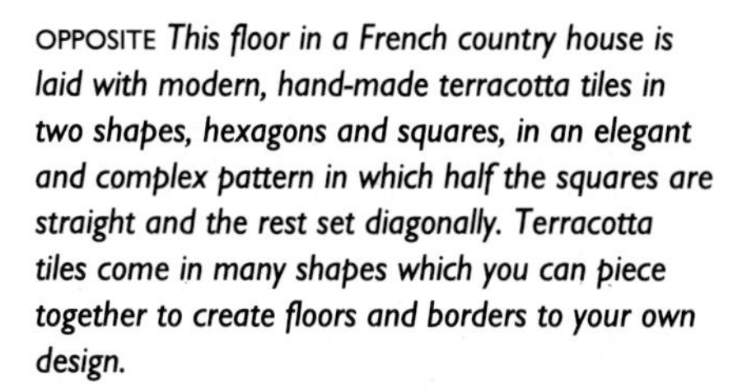

OPPOSITE ***This floor in a French country house is laid with modern, hand-made terracotta tiles in two shapes, hexagons and squares, in an elegant and complex pattern in which half the squares are straight and the rest set diagonally. Terracotta tiles come in many shapes which you can piece together to create floors and borders to your own design.***

Fig. 1

Fig. 2

Fig. 3

Fig. 4

Fig. 5

Fig. 6

Fig. 7

(see Figs. 5 and 7). Squares can also be staggered, rather than being laid in straight rows, and combined with small square inserts in an instep pattern (see Fig. 10). Another alternative is to lay squares in a regular grid but divided from each other by strips of border (see Fig. 11).

A variation on the square tile is the octagon, which either has sides of equal length, or has four long and four short sides. The latter is really a square with the corners lopped off (creating the four short sides) to allow a small square insert of a different colour to be placed diagonally between the larger tiles (see Fig. 8). Octagonal tiles with sides of equal length are laid together in a pattern which resembles a honeycomb (see Fig. 1), or combined with square tiles with sides of equal length to theirs. Diamond-shaped tiles with sides of equal length are like squares pulled out of shape at diagonal corners. They fit together to make whole intricate floors or border strips (see Fig. 2).

Fig. 8

Fig. 9

Fig. 10

Fig. 11

LEFT *Except for the flower at the centre of each star, this Victorian floor could be reproduced using whole and cut modern encaustic tiles. The Victorians used tiles enthusiastically in bathrooms, themselves a new idea, and could choose the component shapes and colours for a pattern from a manufacturer's catalogue.*

Apart from the square, the other most regular shape is the rectangle. This can be laid on its own in a variety of patterns including the classic herringbone (see Fig. 4), at right angles to each other and separated by small squares in a loose basket-weave pattern (see Fig. 3), or in different sizes and combined with squares of different sizes for a more random effect (see Fig. 6). The rectangle has an elegant variation in the six-sided lozenge, in which each end of the rectangle has been extended into a point. These fit together by themselves, in a close-fitting basket-weave pattern, or with squares of various sizes, in a number of satisfying patterns (see Fig. 9). There are other,

OPPOSITE *This ordinary bathroom has been transformed by the use of plain coloured tiles. The floor grabs the attention, then its black and white chequerboard is repeated around the walls and above the mirrors.*

more elaborate, shapes of tile, most notably in the Islamic style, which are generally made of glazed stoneware suitable for walls rather than floors, and described in chapter nine.

Various patterns can be made with plain square tiles of the same size but different colours. They may be different colours because they are made from different materials, or from different types of the same material – red and yellow terracotta, for example. Or they may be manufactured in a choice of colours, like encaustic tiles. Two colours can form stripes in straight lines or diagonals, or the classic chequerboard of alternate dark and light tiles across the whole grid (see illustration on page 86). A double-sized grid can be formed by making alternate squares of four tiles of one colour, using either smaller tiles or larger tiles on a large expanse of floor. The whole chequerboard can be laid straight along the room or passage, or diagonally to the walls. The latter arrangement will make a narrower hall look wider, while the former arrangement will help give an illusion of length to a shorter hall, by leading the eye onwards.

With three colours of tiles, a kaleidoscope of patterns becomes possible, generally with an effect rather like tartan or plaid. Set lines of each colour either straight or diagonally to the walls, or diagonally across a straight grid. Or form a chequerboard from alternate single or double squares (see page 86) of darker-coloured tiles separated by squares of lighter tiles. Alternatively, the squares can be formed from two of each darker colour, set diagonally. Three or more colours can also be placed apparently at random, though it is worthwhile planning the 'random' effect carefully in advance. Do this by drawing it out first, or make a plan using small cut-out squares of different colours, corresponding to the tile colours you are considering (so that you can order roughly the correct number of each colour) and then by laying the tiles out dry and loose on the floor to confirm your design before fixing them in place.

If you are a confident and experienced tiler, or if you are employing a professional tiler, or if you are simply brave and adventurous, you could consider using cut-work tiling. This involves cutting some of your tiles into pieces, usually triangles, either half or a quarter of a tile in size, and rearranging them into a satisfying pattern (see page 87). The edges of each cut piece must be clean and smooth, and the pieces must be fitted back together tightly in their new format, to make the same area as a single uncut tile.

Coloured grouting can be used to help create pattern, but the choice of colours available is generally limited to a few pale shades. Mixing your own with pigment of stronger colour may weaken the chemical mixture and strength of the grout, and you may also find that the colour is streaky because it has not been mixed industrially. Black and dark brown and grey grout are also available as an alternative to white and the pale shades, and are obviously well suited to tiles of darker colour. Whatever the colour, the grout on a tiled floor is likely to weather into a more neutral shade with use.

OPPOSITE *Four different borders made using small marble and clay tiles demonstrate a few of the creative possibilities which a tiled floor offers your home. Draw your ideas on paper first, or use paper cut-outs to create a design.*

LEFT *A glossy, predominantly white tiled floor like this reflects light and makes the room feel fresh and airy, so long as there is plenty of natural or artificial light. In a poorly lit area it might look chilly. The strong border gives the floor a dynamic edge.*

A border gives a tiled floor definition and a finished look. Many styles of tile have a border designed to co-ordinate with them, but it is perfectly possible to create your own border using tiles of contrasting colours and materials. Any shape can be used: large and small squares, the latter possibly in more than one row, diamonds, or tiles of more than one shape (see page 89).

Finally, borders and changes in colour or pattern can be used to manipulate space visually. You can use them to divide a kitchen-cum-dining-room, for example, where you want to differentiate between cooking and eating areas. In other circumstances you might want to unify space rather than divide it. In a home of small proportions, you can lead the eye beyond the confines of one small room by continuing the same tiles through from one room to the next. In a room of annoyingly irregular shape, perhaps one with nooks and crannies, use a colour change or border to create a regular geometric shape, set in from the edge of the room, as a focus in the centre and distraction from the irregularities.

OPPOSITE *A tough tiled floor can withstand any amount of use by hordes of friends, children, grandchildren and animals in rooms designed for frequent use. The same tiles running from one room into another help unify the space and lead the eye onward, here to French windows opening on to a verdant garden.*

CHAPTER NINE

BATHROOMS and KITCHENS

OF ALL THE ROOMS IN A HOME, THE KITCHEN AND BATHROOM ARE THOSE WHERE hygiene is most important and where water is most in evidence. Tiles are ideal in this situation, which is why the Victorians used them so enthusiastically to clad and decorate their new hospitals, food halls, dairies and, in the home, their bathrooms and lavatories. In today's homes, it is the kitchen and bathroom that lend themselves to decorating with tiles more extensively than living-rooms and bedrooms. These tend to be softer places for relaxing and sleeping, whereas the kitchen is the powerhouse of a home.

The status of the kitchen has of course changed out of all recognition since the Victorian heyday of the domestic decorative tile. It is now as much a room for living in as any so-called living-room or sitting-room. The kitchen is where the family gathers in the morning and the evening to eat, but it is much more than a domestic office for the preparation and eating of food. It is a social centre, where members of the household sit and talk or read or play, if they are children, or watch television. Attitudes to food itself have changed too, so that the kitchen is a sort of cultural centre where the culinary ideas and materials of many countries are available, and the interested cook can relax and be creative while assembling fodder for himself or herself, family and friends. As much entertaining now takes place in the kitchen as in the dining-room, a room which in any case is increasingly considered a luxury and even unnecessary.

The bathroom too has changed, from being purely utilitarian, and often cold and uncomfortable, to being a real room, interesting and welcoming and a pleasure to spend time in. The effect of these changes is apparent in the amount of time and effort we put into devising decorative schemes for our bathrooms and kitchens. As with floor tiles, decorated and plain tiles are now available from all over the world, combining the necessary hygienic qualities (which have never gone out of fashion) with beauty.

The importance of hygiene in kitchen and bathroom means that you or your decorator need to consider carefully the type of adhesive and, even more important, the grout to be used. The adhesive needs to be water resistant. There are various

OPPOSITE *Simple blue and white tiles have here been used in a flamboyant design which stretches from bath to ceiling, as well as covering the side panel of the bath. The strong pattern has a refreshing, random feel and offers strong but pleasingly cool balance to the bright, hot colours elsewhere in the room.*

RIGHT *In a small bathroom, coloured tiles have been used sparingly for maximum effect. Small dark tiles have been placed in squares of four between larger white tiles, but around the window the diagonal angle is suddenly abandoned. This change draws the eye to the window and beyond, distracting it from the small size of the bathroom.*

types of grout available, such as those recommended for use on kitchen walls and in bathrooms and showers (with added fungicide) and others for kitchen work surfaces. The latter type is an inert compound which will not absorb food particles or bacteria – anything that lands on it can be completely washed off. Check that the one you buy has passed tests which allow it to be used in food preparation areas in locations like restaurants and canteens where the general public could be affected by a drop in hygiene standards.

One element of decoration which many people consider unsuitably unhygienic in a kitchen, and to a lesser extent in a bathroom, is window dressing. Curtains absorb moisture and grease, and can be a fire hazard. You may decide to do without and instead use a plain blind or shutters to retain privacy at night, or you may not feel the

LEFT *The floor colours here are fresh and jaunty, attracting the eye. Tiles have been cleverly used to differentiate between the more important part of the kitchen, where the table is, and the annex, where the fridge can be seen. Both areas have the same sky blue and white chequerboard with a blue border, set at the same diagonal angle, but the tiles in the important area are twice the size.*

need to cover your windows at all, if you are not overlooked. In any of these cases, consider framing the window with coloured or patterned tiles. Make sure that any cut tiles necessary for a perfect fit are at the corners of the window, not in the middle, or position the frame of tiles an inch or two away from each side of the window to avoid completely the need for any cut tiles.

Kitchens and bathrooms are the rooms in which to use picture tiles. Here they will be properly seen and enjoyed – most are in any case unsuited to the heavier wear floor tiles suffer. Picture tiles can be used *en masse* for a richly ornamented effect which the Portuguese would recognize from their great tradition of cladding whole walls and rooms with a riot of pattern and colour; or used in panels and strips to accent particular parts or characteristics of a room; or dotted about with plain white or coloured tiles between, which allows the picture on each to stand out and be seen more clearly. Ideal images for kitchens include fresh vegetables and fruit, flowers, ornamental birds, and words like the names of herbs and Italian or French cooking terms or words for foods. Unless you are vegetarian, fish are another suitable subject; but even if you are a keen carnivore, you might consider pastoral pictures of cows and sheep to be tasteless in this context.

Before deciding on any set of tiles, however, take a step back and look at your kitchen as a whole. What is its style or mood, and what are the dominant colours? Do you have a solid-fuel stove with a coloured enamel finish, or cast-iron enamel-coated pans? What colour are your storage jars? What are the colours and style of the pattern on your china? Can the units be painted, in order to create a fresh new colour scheme and co-ordinate with the tiles? If this seems too drastic, is it possible to buy replacement china knobs for drawers and cupboards in a colour which links the units to the ceramic tiles? An alternative to the latter plan is to paint or stain wooden knobs, and then protect them with several coats of clear varnish.

Besides the floor, other obvious places in which to put tiles in a kitchen are on the wall above the work surface, and face-up on the counter, forming the work surface itself. On the wall above the work surface, tiles provide a practical, wipe-clean background for potentially splashy activities like beating, blending and mixing. Space on the wall is often limited by wall-mounted cupboards, so fewer tiles are needed to make an impact. A border along the top of the tiles here is less likely to be obscured by storage jars or objects such as small electric appliances than one along the bottom. Borders along both top and bottom create a strong outline and horizontal emphasis which would be especially refreshing in a very small kitchen.

If there are no wall cupboards, a splashback of tiles can be continued up the wall to the ceiling, or it can be limited to a height of four or five tiles, or it can end at any point in between. Considerations are the height of the room and the use to which the wall space is to be put. Shelves for objects or books, pan or drying racks, pictures . . . a low splashback beneath any of these will make sufficient visual impact and be as practical as a taller one.

ABOVE *Ceramic tiles have been used for centuries as an attractive and hygienic solution to decorating areas of the kitchen where food is cooking and there are likely to be splashes on the floor or walls.*

OPPOSITE *The bold floor in this kitchen adds a third layer of geometric interest. The first, or highest, is the windows with their unusual triple division in the upper half. The next is the cupboard doors, which have square air holes and square handles. Last comes the floor, which is a classic chequerboard.*

ABOVE *When using coloured tiles in a kitchen, remember to consider the colours of your china and pans. Linking them can be very effective.*

If you want to use just a few tiles for maximum impact, limit them to places in the kitchen where they will be of most practical use, such as behind the sink or cooker. A panel of tiles consisting of a border framing a picture of flowers in a vase, a colourful exotic bird or a figure in peasant costume, for instance, would be ideal here. So would a patchwork of tiles with the same finish but many different colours, or many shades of one or two colours. If you have collected some Victorian or other antique tiles, and if they fit in with the style and colours of your kitchen, this might be a good way of displaying them, where you will see them regularly and gain maximum pleasure from them, at the same time putting them to practical use. As elsewhere, the colour to use most carefully is white. Remember that in a kitchen it will show splashes of dirty water or grease more readily than other colours. Some people consider this a real advantage in hygiene terms since you can easily see what needs cleaning.

Other possible places to use small quantities of decorative tiles are across the front of a cooker hood, and along the plinth under your units. In the latter case, your kitchen really needs to be of the farmhouse rather than the galley type and size, as the tiles will only be seen properly from several feet away. Plinth tiles also need to be quite tough as they will tend to be kicked in this position. The usual arrangement for the plinth is to tile it with the same tiles as those used on the floor, but you do not have to stick rigidly to this rule, just as you do not have to limit terracotta and other tiles used for flooring to the floor underfoot. You can just as successfully use terracotta or marble for the walls above your work surfaces, where you can enjoy their subtle, swirling colours. Terracotta will have to be thoroughly sealed; thinner, machine-manufactured tiles may be more practical than thicker, heavier, hand-made tiles as the former are smoother and therefore more hygienic, and their lighter weight will cause fewer problems when attaching them to the wall.

Tiles on work surfaces are a cause of some controversy. Most people whose kitchen counter-top is constructed of tiles are very happy with them, but some people regret the choice, not on aesthetic grounds but because of hygiene and practicality. They find the overall surface is not regular and smooth enough to work on easily, and the grouting is never completely level with the top of the tiles, which means that liquids and food particles inevitably get trapped. This is why many people who consider having a tiled work surface decide against it, and choose instead a material such as wood, marble, Formica or stainless steel. However, you need not do without tiles completely if you choose one of these surfaces: a ceramic border or lip can be used to finish it off along the front edge. The same is true in a bathroom, where you may prefer to have basins and bath set into a smooth, wipe-dry surface edged with tiles, rather than a surface made of tiles. Hygiene on bathroom surfaces is not perhaps such a vital issue, however, since you do not eat food prepared on them.

The bathroom is the room where your tile ideas can run riot. It is possible to go to the extreme of actually having your bath and basin lined with tiles. The human

body is rounded, however, and a tiled bath has compulsorily flat surfaces, so be prepared for a little discomfort as the price to be paid for such glorious luxury. Both bath and basin will have to be carefully laid to ensure that the water all runs towards the plug hole. The more usual use of tiles in the bathroom is on the floor and walls, especially walls immediately above bath and basin. A fully-tiled bathroom is intensely practical, since this is regularly the steamiest room in the house as well as the wettest. If you have a shower, either in the bath or as a separate unit, this is an area to consider tiling up the whole wall, even if you do not do this elsewhere in the bathroom. A separate shower made into a cubicle, or an alcove around a bath built with partition walling, will look, and be, more solid if the partition walls are tiled, and of course the tiles provide an extremely efficient protection against damp penetration.

ABOVE *This bathroom wall has been decoratively tiled, and the basin, which is a spectacular piece of design in itself, has been set against it afterwards. Soap and accessories like the tooth mug and towels have been chosen to match the colours of the tiles.*

RIGHT *The colour of this art deco bath and basin have been picked out on the wall with a border of diagonally placed green tiles which runs around the room just above the height of the basin taps. The rest of the bathroom has been dramatically tiled in black.*

Completely tiled walls are most luxurious, and are potentially satisfying in visual terms too, since the unity of the walls draws together what can otherwise look like nothing more than a collection of peculiar china shapes (the basin, lavatory, bath and so on). The next most unifying arrangement is to tile all walls up to a certain point above the china, finishing this off with a border. And on the floor, a regular tiled shape like a square or rectangle (or circle, though this is much harder to lay), set inside the china fittings, will also have a unifying effect. It helps too if your china is all the same colour and matches in style. Colours and patterns on the tiles obviously need to be compatible with the china. Now that colours like avocado and burgundy are mercifully out of fashion for bathroom china, and the most popular colour is white, this is unlikely to be a problem. This leaves you free to set a mood in your bathroom (and kitchen) with the colour scheme you choose for your tiles.

Colour can be used to emphasize aspects of space and mood. A small pattern in gentle colours will be soothing; a larger pattern in stronger shades, perhaps a chequerboard of two or more colours, will make your bathroom a wakeful and invigorating place to be. Horizontal or vertical emphasis can be given to a bathroom with stripes created by tiles of alternate colours going along or up the walls, or with tiles that have a pattern with strong emphasis in one direction or the other. An area of solid colour will contrast with an area of patterned tiles and throw the pattern into relief.

OPPOSITE *Some bathrooms are places for quiet and repose; this one is likely to make you feel refreshed and energetic even before you've had a shower. The dynamic chequerboard of blue-green and white on the walls is relieved by panels of plain white which draw and rest the eye in the same way as a window.*

RIGHT *The spartan whiteness of this elegant bathroom is relieved by a slender black and grey tiled border leading the eye around the room. Below this, floor and walls are covered in large plain white tiles which create an atmosphere of cool serenity.*

A border can be placed at any height you choose, depending on where you want the visual emphasis. At the lowest level, tiles can be used instead of wooden skirting or base boards adjoining the floor, which is where the Dutch placed them when they began to use tiles on walls in the sixteenth century. This is effective in a large bathroom with expanses of wall with no china against them – less so in a small room where the line is frequently interrupted. Just above bath or basin height (the basin is higher so the line will not be interrupted if the border is above it) is another option. The wall tiling can stop here or continue up the wall above the border.

The next obvious punctuation point up the wall is the height of a low ceiling – seven or eight feet. A border here can help give an illusion of height if you have a low ceiling, and will give the eye a resting point if the ceiling soars away above it. Alternatively, in a low-ceilinged room, place the border at one of the levels below head height in order to distract from the lack of a soaring ceiling. A tile border need not go the whole way around a room if only one area, around the bath perhaps, is tiled to the top: the line can be continued by a dado rail, picture rail or a painted or paper border.

Many bathrooms and lavatories are small, in which case use either warm colours to create a sense of intimacy, or cool colours and plenty of mirror to give an impression of greater space. In a cool bathroom the lighting, especially electric lighting, must be well-placed and plentiful or the effect will be cold and dingy. Whether you use patterned tiles, one or two colours of plain tiles, or a mixture, it is worth considering your basic colour combinations. Some fresh, cool combinations are blue and white (always popular, in any room), black and white, grey and white, cream and green, green and blue, green and black. Warm combinations include rich yellow with white or cream, rust and coral, pale and dark pinks and reds. When choosing colours, remember to take your towels into account, and the colour of the bedroom out of which the bathroom opens, if that is the arrangement of rooms.

In cities especially, many bathrooms are not only small but internal, that is to say without a window. A tile panel showing birds or flowers or an outdoor scene – a lush landscape perhaps, or a seascape with boats – will act as a focus in the same way that a window does, while leading the imagination out to wider spaces and thus introducing a sense of fresh air and the outdoors. This will naturally diminish any feeling of enclosure caused by the lack of a window. Whether or not your bathroom has a window, consider placing your tile panel (or panels) opposite a mirror or mirrored wall, to multiply its effect.

Other figurative subjects appropriate for individual picture tiles on bathroom walls are watery ones like fish, fishing, shells, the sea, the seaside and boats. Verdant subjects which will make your bathroom feel like a cool shady bower, such as abundant flowers and swags and garlands of flowers and foliage, are appropriate too. For a baroque effect, use more than one pattern in your bathroom – one pattern, denser and more elaborate, on the side of the bath, for example, and a simpler version of the same on the walls above the bath and round the basin. So long as the patterns and their colours relate closely to each other, the effect will be rich and sophisticated.

Modern life can be exciting but stressful. For many people, especially those whose home teems with children and animals, the bathroom is almost the only place in the house where they can close and bolt the door without being antisocial or feeling irresponsible, and relax completely. A mug of tea or coffee at the beginning of the day or a glass of wine in the evening, partnered by a paperback book (no tragedy if it gets wet), transforms the bathroom into a haven of self-indulgence. Under these circumstances, you may feel your bathroom deserves really special treatment. This could be a large and spectacular tiled panel alongside the bath or on another wall where you can look at it from the bath, or a specially commissioned mural. If you intend to live in the house for a long time, the expense of a mural will be justified. The next chapter, Walls and Decorative Details, gives more information about commissioning special work from an art or craft tiler.

CHAPTER TEN

WALLS and DECORATIVE DETAILS

FLOORS, KITCHENS AND BATHROOMS ARE UNDOUBTEDLY THE PARTS OF THE INSIDE of a home which lend themselves most obviously to being tiled. But they are by no means the only places where tiles can be used to great visual and practical effect. Other possibilities fall roughly into two categories: walls, and decorative details. In the case of walls, it is not necessary to go as far as the Portuguese, who have a tradition of lining whole rooms with a riot of pattern and colour, and still do not hesitate to decorate in this manner today. A considerable impact can be achieved with far fewer tiles, though the blanket coverage approach is undoubtedly dramatic, and can be stunning.

Two ways of covering a wall but not the whole room with tiles, are the mural and the dado. Both are ideal in areas which may come into contact with water or be subjected to the enthusiastic attentions of animals or children. The walls of a narrow hall, for example, where in winter or rainy weather (of which there is plenty in Britain especially) wet umbrellas, rain-soaked outdoor clothes, and snowy or muddy boots are deposited, would be protected by a tiled dado.

The dado is the bottom section of the wall, below a line called the dado or chair rail which is generally around the height of the back of a dining-chair. A dado rail may be higher. The Victorians often decorated the dado with much tougher materials than other parts of the wall, in recognition of the extra wear and tear it suffers. They covered it with thick, varnished, marbled paper, or an embossed, patterned material rather like linoleum, or, of course, glazed tiles. Instead of becoming worn and marked, the tiled hall could regain its freshness with a mere wipe of a clean cloth. A tiled dado need not of course be confined to a narrow hall. It would be appropriate in any entrance area at the front or back of a house, or in a cloakroom or lavatory, especially in a house or mansion block built in the Victorian or Edwardian period or style.

A traditional tiled dado consists of several parts: the dado rail is a pronounced ridge; the wall below is covered with a combination of plain coloured tiles, decorated tiles and slips – narrow strips of plain colour which act as spacers between one element and another; and at the bottom is the ceramic skirting, which again protrudes

Tiles in a chequerboard of two bright colours, one darker and one lighter, create a practical and cheerful covering for the lower part of a wall.

OPPOSITE *The dynamic tiled surround of this fireplace, with its multicolour zig-zags, is impressive in itself as well as providing a visual link with the surround of the stove in the room beyond, seen through the doorway.*

ABOVE *A frieze of plain tiles arranged in a geometric pattern below the picture rail ties in well with the fabulous primitive mask hanging below.*

and protects the last few inches of the wall in the same way as a wooden skirting or base board. A less traditional dado of tiles is simply the lower section of wall covered with tiles. Visually it is a good idea to give your dado a definite top and bottom, perhaps with borders of darker or patterned tiles. Alternatively, the tiles on the body of the dado could be set diagonally, creating a trellis effect, with the tiles at top and bottom running parallel with the floor.

For a real ceramic dado complete with rail and skirting, it would have been necessary a few years ago to go to a specialist pottery concerned with conservation work, and even then the job might have been an expensive special commission. No such drastic measures are necessary now, as the manufacture of all the tiles for a full dado has been revived by at least one factory, the historic Jackfield tile-works at Ironbridge, Staffordshire, in England, which had its heyday in the reign of Queen Victoria. Using Victorian machinery and methods of manufacture, Jackfield has succeeded in recreating a glowing depth of colour which makes some modern tiles look insubstantial by comparison.

The most modern tiled decoration, by contrast, is a panel or mural created especially for you. There are many tile artists who will take on special commissions for anything from a panel of four or six tiles to a mural that covers yards and yards of wall. Some also offer a complete house service, advising you on tiles for all the rooms where you had envisaged them, so that your entire home will have a consistent style.

The difficulty is often in finding tile artists whose work you like. Ideally they should be near you geographically so that you can meet them, see other work they have done and liaise closely with them as your design develops. Word of mouth and personal recommendation is one way of discovering a suitable tile maker. Art and craft organizations near you may also be able to provide some information, and national organizations are usually keen to put you in touch with their members, who may get work as a result. Another approach is to look in the classified advertisements in interiors magazines, and in art magazines.

Once you have a list of tilers, telephone them all and ask for illustrated literature about what they do, or visit them in person to see for yourself. Certainly visit a shortlist of candidates after your initial research. Do not be afraid to ask questions, including technical ones. Craftsmen tilers mostly enjoy the educative process and find that receptive customers quickly become enthusiastic about the whole tile-making business. If possible, be clear about what you do, or do not, want; and be guided by instinct if you feel the tiler is not listening properly but simply promoting his or her own ideas.

If you are lucky enough to have an indoor swimming pool, this would be an ideal place to have a mural. It could either show a figurative scene – anything from a seascape with boats to a lush green landscape – or a fantasy such as the underwater world of Neptune and his followers, or it could be a dynamic abstract design of shapes and colours. Think hard about what you really want; if possible look at other murals, and choose a tile artist whose work is compatible with the style of mural you

LEFT *A hard wearing and practical Victorian style tile-covered dado, complete with chair or dado rail at the top and skirting or base board at the bottom, all formed from glazed tiles. A row of decorated tiles has been included below the chair rail to lighten the effect of rich, dark colours.*

ABOVE *Decorated tiles around the edge of the pool and a mosaic of lively intertwined fish by Odette Wells and Sebastian John of Tiles of Stow.*

OPPOSITE *Terracotta tiles need not be confined to the floor. A fire surround is another spot where the tile's practical nature is put to excellent use. Here, floor tiles have been continued up the wall to clad the entire chimney breast, forming a warm frame for the blazing fire. The interesting pointed arch is edged with brick.*

envisage. If you commission a smaller mural, consider having it mounted on a panel and fixing this to the wall, rather than attaching the tiles directly. In this way you will be able to take it with you, if or when you move house. And do not forget that tile makers will usually be happy to make you tiles for use anywhere in your home, not only in decorative panels.

Fire surrounds and fireplaces are always a focus of attention in a room, and provide the ideal setting for a small collection of tiles. Larger numbers can be used to face the inside of an inglenook fireplace, if you have one. In summer, when there is no fire in the grate, the tiles can provide alternative interest, whatever the size of your fireplace. In winter, people instinctively gather round a crackling fire, for the warmth and because its noise, light and movement draws them to it. Contemporary speculative house builders are once again installing fireplaces after decades of believing them redundant, because they have realized that their customers like the feeling of cosiness and welcome an open fire gives to a room.

If you are installing a fireplace in the Victorian style, with a cast-iron surround including mantelshelf and basket, and a panel of tiles inset down each side, consider the colours and patterns already in the room before choosing. Such fireplaces generally have the tiles already installed, but it may be possible to make your own choice of tiles. In either case, you could construct a tiled hearth to tie in with, but not compete with, the colours and pattern of the fireplace tiles. If the latter have a bold blue and white floral pattern, for example, your hearth could consist of tiles which are predominantly white with a small motif in blue in the corners and perhaps in the middle.

Some of the most interesting Victorian fireplaces have panels which contain not rows of patterned tiles but tile picture panels. The five or six tiles on each side together make up a complete picture with subjects like vases of flowers or personifications of Peace and Plenty. If you are constructing a modern fireplace, or filling the surround within a wooden chimney piece, you could follow the same idea. You could commission a tile artist to decorate panels or a complete surround of tiles with a subject relevant to you – even a portrait of your home and family. Alternatively, there are so many decorative tiles already on the market, from bright, hand-painted peasant creations to elegant tiles in the blue and white Dutch style, that there should be no difficulty in creating a fire surround that exactly matches your taste and your room.

An unusual and rare form of Victorian cast-iron fire surround had detachable metal 'wings' each side of the grate, each containing a tile panel. The purpose of this was to enable you to change the tiles according to the season, or indeed mood. You would be extremely lucky to find a genuine Victorian fireplace of this type today, but it might be possible to commission a craftsman blacksmith to create something similar if the idea really appeals to you.

Two practical points should be considered when choosing and installing tiles for a fire surround. Check the heat resistance of the tiles you wish to use with your supplier. And if you are using tiles with a diagonal design, be careful to attach these with

SIMPLE PAINTED

RIGHT *A detail of the fire surround on page 104, showing how small tiles of triangular, square and rectangular shapes have been assembled. The careful choice of colours creates variety without the patterns looking a mess, and the soft grey grout unifies the entire arrangement.*

the pattern going the same way on them all. It is astonishingly easy to slip in the odd tile going in the wrong direction, especially near an edge or the bottom. This is equally true of course for any panel of tiles, be it in a fireplace, above a kitchen work surface, alongside the bath or in any other room.

If a joiner or builder is making your chimney piece to your own requirements or design, you could ask for a mantelshelf of sufficient thickness to take a narrow border of tiles along the front – either a plain-coloured border or one whose pattern complements the pattern and colour of the tiles in the surround beneath. Another way of using tiles on a chimney-piece is to range them along the back of the mantelshelf, where they will be nearly at eye level. Whether or not this arrangement will appeal to you depends to some extent on what you intend to stand on your mantelpiece. If you have a collection of china or glass that you plan to display there, tiles are an ideal backdrop. Engraved glass will come to life in front of a frieze of plain, dark-coloured tiles, and patterned china can be provided with a backdrop of tiles of sympathetic pattern and colour. If, however, your mantelshelf tends to become home to a

LEFT *The splashback behind a basin is tiled with small, roughly finished marble tiles, some etched with primitive figures of animals and men. The mount for the taps is wooden, the taps are brass, and a mound of pebbles and shells completes the assembly of natural and covetable materials.*

picturesque clutter of postcards, invitations, shells and stones, family photographs and candlesticks which will completely obscure the frieze of tiles, you might feel they are wasted there.

There are many other places in a home where a small number of tiles can be used to great practical and decorative effect. Bedroom hand basins generally need a splashback to protect the wall behind, for which a small panel of decorative tiles is ideal. A basin in this setting wants either to melt as unobtrusively as possible into the room's decorations or to be made an object of interest. You are sure to find that tiles fulfil either of these requirements.

One simple way to make an interesting splashback is with a panel of plain, glazed tiles, about the same width as the basin, edged all round with a patterned border. The plain tiles in the centre could be the same colour as the basin china. Depending on how much water you imagine the basin users will splash about – which may be plenty, if they are children – you could also cover the area of floor immediately beneath a pedestal or bracket basin with tiles. This 'mat' of tiles needs to be several

RIGHT *A tile frieze and the frosted glass in the window have the same decoration of stylized art nouveau flowers. The tiles could be commissioned to match existing glass, or the glass decorated to match the tiles.*

inches wider and deeper than the basin, and stretch back to the wall, to be really effective in practical terms.

Indoor windowsills decorated with tiles look unusual and attractive, especially if your curtains are unfussy or if you prefer a blind or shutters to window dressing or drapes. Make sure that you start with a whole tile in the middle at the front, or room edge, of the sill. Any pieces of tile should be at the sides and back. In the same way, a whole window embrasure can be lined with tiles, which will help reflect light into the room. As described in chapter nine, Bathrooms and Kitchens, a whole window can be framed with tiles. This is easiest to do if the window is small, but larger tiles can be used to frame bigger windows and even doors. A doorway with a surround or architrave which is an arch of patterned tiles becomes a dramatic entrance or exit, attracting the eye and leading it on to the room or space that it frames.

Furniture can be built around tiles, or embellished with them, or given a new lease of life. For centuries, furniture has been designed around panels of ceramic decoration. The Victorians, and especially the Arts and Crafts artists, were fond of chairs, sideboards, tables and other furniture which incorporated decorated tiles. Such pieces are still to be found today, or a contemporary craftsman can design and make you your own. Alternatively, raid your attic or cellar, or keep a look out in junk shops and at auctions for furniture which needs attention and which could be made more beautiful or more useful with the addition of tiles. An old blanket box, for instance, with a cracked and broken lid, can be refurbished with a coat of paint and a tiled top which will transform it into a useful and tough table. A worn tray or small table will be made beautiful again if topped with tiles, though in the case of the tray you need to be satisfied that the tiles are heatproof and not too heavy to make it impractical.

For a really dramatic kitchen or dining-room table, mount a complete panel of patterned or plain tiles on to a board or an old door and provide the top with sturdy legs at the corners, and possibly along the side, depending on the length. Tiles are heavy, especially craft and hand-made ones, and you should be satisfied that they are properly supported. Alternatively, commission a joiner or cabinetmaker to construct a table on which to mount tiles.

Mirrors offer an ideal opportunity for decorating with tiles. Create a frame from border tiles for a smaller mirror, and from whole tiles for a larger mirror, perhaps a full-length one. Finally, single decorated antique and modern tiles can be used and displayed to great effect. Back single ones with felt and use them as stands for hot pots or the teapot at table. Display single tiles on the wall or on a shelf, mounted on a suitable hanging spring or wooden or perspex stand. They will give you pleasure every time you pass by.

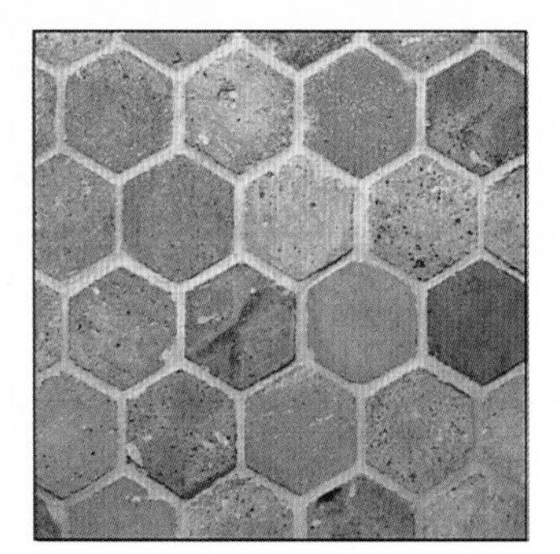

CHAPTER ELEVEN

OUTDOORS

ONE OF THE JOYS OF HAVING A GARDEN IS BEING ABLE TO USE IT AS AN EXTRA space for living in when the weather allows. Meals eaten outdoors in the summer sunshine, quietly by oneself or in the company of a crowd of friends and children, have a romantic, Mediterranean aura. Simple food seems special and Sunday lunch a feast. A flat, hard area in the garden is ideal for a table and chairs, which might otherwise sink into the ground and wobble lopsidedly. Another practical consideration not to be dismissed concerns mowing the grass. If your table and chairs are in a grassy area, the person mowing has to move them away and back again, a chore he or she will find a considerable nuisance. This is especially true in high summer, when the grass needs cutting almost every day. The problem is avoided by having a paved area for eating on.

The flat hard area can be near the house, which is practical for carrying things in and out; in a larger garden you could put it in a pretty spot where there are views and shade in case of hot sun. Next to the house, a terrace is an extension of your kitchen or sitting-room – you can slip out to sit and read your newspaper or prepare vegetables at the slightest sign of sunshine and warm weather. If you have small children, such an area is ideal for a sandpit, whose inevitable overflow can easily be swept away.

If your garden is small, perhaps not more than a tiny back yard or a roof terrace in a city, the only sensible way to treat it is to pave it. A postage stamp of grass is not practical (apart from anything else, where do you keep the mower?) and can look silly. Paved areas need minimal maintenance and give a solid surface for terracotta pots and other containers in which you can grow anything from vegetables to small trees. Tiles will create a visual contrast with other structural materials in the garden, such as the stone or brick edges of raised beds, and the walls that rise behind them to create the garden's perimeter and give it height.

The materials that usually spring to mind first when planning the bones of a garden are the local stone, flagstones, bricks and gravel. In colder parts of the world, tiles are generally thought of as an exclusively indoor material. Indeed, it is true that they

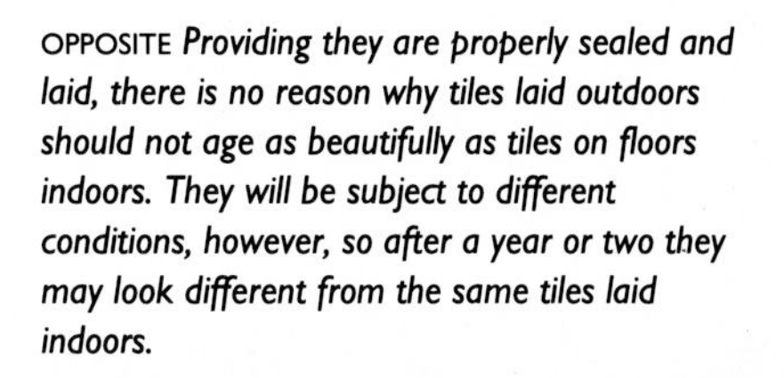
OPPOSITE *Providing they are properly sealed and laid, there is no reason why tiles laid outdoors should not age as beautifully as tiles on floors indoors. They will be subject to different conditions, however, so after a year or two they may look different from the same tiles laid indoors.*

are not suited to great extremes of temperature. But in a temperate climate like Britain's, almost any tile suitable for indoor floors can effectively be laid outdoors, adding hugely to the repertoire of available paving materials. You should be aware, however, that over the years tiles laid in your garden are bound to take on a different look to the same tiles indoors, because they will have experienced different conditions. As long as they are properly laid and maintained they should last equally well.

Apart from the climate, practical details which will affect the suitability of tiles as opposed to other paving materials include their position in the garden, and pollution. Inner-city areas with heavy pollution may not be ideal, especially with porous tiles like terracotta, and neither will a position which is overhung with vegetation and tends to be damp. The tiles will almost certainly be affected by algae and become green and slippery. The ideal spot is somewhere dry and sunny.

Terracotta, slate and stoneware tiles are those best suited to outdoor use, and should be treated with a top quality waterproof protective sealant on top of the sealing treatment you would give them if you were laying them indoors. The waterproof sealant will be absorbed and block off the pores of the material, forming a skin over it. Once you have done this, surface abrasion is the only maltreatment that may be capable of damaging the tiles – spills should have no serious effect. Every other year, when the warmer, drier weather arrives in spring, give the tiles a thorough scrub with a mild acid and when dry add another coat of waterproof protective sealant. Mild acids include so-called 'patio cleaners' which you can buy in garden centres, and any other very mild solution of phosphoric or hydrochloric acid. These will clean dirt, kill algae and cure minor efflorescence.

These materials – terracotta, slate and stoneware – do not have to be used exclusively of each other. You can design your terrace in a pattern using any of these together, and also natural stone, to create changes in colour and texture. Glazed tiles can also be used outdoors, but their use is more problematical. The most beautiful glazed tiles are often those made by hand, which tend to be deep and chunky, with a thick glaze which gives their colour added depth and vibrancy. But of all glazed tiles, these are least suited to outdoor use. If moisture penetrates the depth of porous clay beneath the glaze, changes in temperature will cause the tile to expand and contract, which in turn causes cracking on the surface.

Machine-made glazed tiles are more likely to withstand outdoor use than hand-made glazed tiles, but they are still not ideal. Thinner, and with a thinner, vitrified glaze, their colour is flatter and duller. However, they are less susceptible to moisture penetration and surface damage, and the duller glaze is less likely to be slippery. Hand-made glazed tiles can be hazardous if their surface gets wet. It is perhaps better to use glazed tiles sparingly, as a highlight adding colour and interest to an outdoor floor, or to create a striking border, rather than as a general paving material.

Areas of paving, paths and paved steps give a garden of any size, but especially larger gardens, visual variety. They bring crispness and an edge, adding definition to

OPPOSITE *Tiles are used outdoors more as a matter of course in hot countries, as this picture shows, but are perfectly able to cope with a temperate climate. On a balcony, outdoor steps or in a small or enclosed space like this courtyard, they are an ideal ground cover and surface on which to stand plants in containers.*

the softness and abundance of lawns, flowers and foliage. At the edge of lawns and grassy areas, where these abut flowerbeds, tiles serve a more practical purpose than less regular paving materials. A line of tiles along the edge of the grass makes it easier to mow neatly, and makes it correspondingly less likely that plants trailing over the edge of the bed will be chopped back or otherwise damaged in the mowing process.

In the same way, tiles along the edges of a gravel path help contain the stones as well as adding visual interest and definition to both path and the beds or grass alongside. A path made entirely from tiles can be charmingly patterned between straight tiled edges. Best of all is a network of tiled paths, between the beds in a rose garden, a potager, herb or kitchen garden, for example. The whole garden can be designed with an overview which allows for the fact that some paths are more important and wider than others, the others being byways off the main axes and vistas. Different tiles can be used for different categories of path, or the same tiles in different sizes, or tiles made in the same material but in different shapes.

A garden on different levels has an inbuilt advantage over a flat one, in terms of visual interest. Steps leading the eye from one level to another are a focus in themselves as well as an introduction to a new view or vista. Narrow steps are mysterious, almost secretive; broad steps which allow two or more people to walk abreast up and down them are generous and welcoming, even romantic. In either case, if there is any sign of the treads of steps faced with tiles becoming slippery and dangerous, scrub and reseal the steps each year rather than every other year.

As with an indoor windowsill or work surface, make sure when laying tiles on steps that the front edge of each step is constructed from whole tiles, with any cut pieces at the back; and that the tiles are laid symmetrically with either a whole tile, or the joint between two whole tiles, in the middle of the step. If an exact number of tiles does not fill the width of the step, fill the edges with pieces of tile of equal width, rather than filling one side only and distorting the symmetry. The same applies to the risers, but here you need not be so conscious of practicality. Because they will not be walked upon, you need not worry about these tiles being slippery, so you can include decorated and coloured glazed tiles for consideration when planning the steps' appearance.

Risers are seen in a way that treads are not, and they therefore offer an unrivalled opportunity for decorative detailing. You could decorate the risers with patterned tiles, or pictorial ones, or plain-coloured ones, or letter tiles spelling out names and dates, or even a favourite poetic quotation. For neatness and precision, plan the height of each step to accommodate an exact number of the tiles you want to use. The structure of the steps beneath, to which tiles on risers and treads are attached, must be solid and secure, so that no undue pressure is placed on the tiles when people walk up and down the steps. Ideally, the underlying structure should be made from concrete or concrete blocks cemented together. The tiles should then be firmly fixed and grouted in the same way as an outdoor tiled floor. Work from the bottom

LEFT *Different arrangements of tile have been used in this garden to indicate different uses for the paved areas. The path has tiles laid diagonally with small patterned inserts; the area leading to the pool is laid in a simple grid.*

to the top of the steps, from front to back of each tread, and tile each riser before progressing up to begin the next tread.

An outdoor tiled floor, forming a terrace or seating area in your garden, should be designed and planned in much the same way as an indoor floor. The same huge range of patterns is available to you, including the classic chequerboard of alternating dark and light tiles, and the variety of terracotta shapes such as hexagonals, octagonals and lozenges. But because an outdoor tiled terrace is only used for part of the year and does not have to accommodate a collection of furniture and rugs, you can perhaps allow yourself to indulge in ideas that are a little wilder than you would use indoors. Borders can be broad and strong, the pattern within can be bolder and more daring. You could draw a maze with tiles of different shades or colours, for instance, or spell out initials or a monogram.

If possible, plan the area as a regular shape – a square or rectangle – and if it is irregular, lop off the odd angles which prevent it from being regular, for the purpose of discovering where the main body or grid of the floor will be. Peg taut strings into the ground along your axes, and be sure that these give you the visual lines you want. When calculating the position of your first tile, allow for the same width of grout joints between tiles as you would indoors.

Tiles outdoors must be laid with meticulous care to ensure that there is no possibility of moisture being trapped underneath. This will shorten the life of the tiles by swelling in icy weather and causing cracking. Tiles should never be laid straight on to soil or sand. Dig the area out to a depth of 15cm (6 in) and fill it with rubble or hardcore which will ensure efficient drainage underneath the paving. Pack this down thoroughly, lay a generous screed of concrete, and bed the tiles while the concrete is wet. Finally, grout with a product which is suitable for outdoor use.

Besides sealing the tiles thoroughly, you could take the extra precaution of sealing them individually before you lay them. It is also possible to seal the bottoms, to protect against moisture penetration, but this should not be necessary if the tiles are properly laid. The ideal weather in which to create your outdoor tiled area is obviously warm and dry. Rain will ruin the process. You should also be certain that the tiles themselves are completely dry before beginning. Water is an important consideration, however, and must be taken into account. When it rains, where will the water go? You do not want it to lie or be trapped on the paving. Use a spirit-level to slope your tiles very slightly towards a drain or gully or other place such as a well-drained flowerbed where the water will be able to escape or may even be useful.

The conservatory is an ever more popular addition to modern homes: a room which allows you to enjoy the outdoors even when the weather forces you to stay indoors, but which opens out to become part of the garden when the weather is fine. Tiles provide the ideal flooring in a conservatory. The floor should be planned taking into account the style and function of the room out of which it opens, the type of garden it adjoins, and the size and style of the conservatory itself. Otherwise the floor should be planned and laid in the same way as an indoor floor, giving special consideration to the focal point where you want to see whole tiles, not pieces.

OPPOSITE *Conservatories tend to become part of the garden in summer, with people coming and going from outdoors. Tiles provide a suitably rugged flooring as well as being warm and decorative in winter. Octagonal encaustic tiles have been laid here with a border and white inserts, except for two lines with black inserts.*

CHAPTER TWELVE

PRACTICALITIES

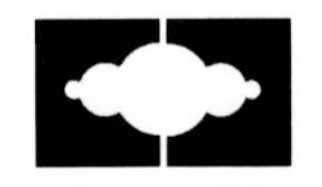

THIS CHAPTER IS AN OUTLINE OF THE BASIC TECHNIQUES INVOLVED IN LAYING AND fixing tiles. It may make tiling seem impossibly difficult and problematic, but in the long run it really is worthwhile doing the job properly. After all, a well-laid tile floor or wall will last a long time. And you will probably find not only that it is not as complicated as you thought, but also that you get better and quicker at it as you go along. Begin your tiling career in a spot which is not the most visible, and by the time you get to the showpiece area in the middle of the floor or the bathroom wall, you will almost be an expert. Alternatively, if the prospect of doing your own tiling gives you sleepless nights, get a quotation for the work from a professional. The cost of getting a professional to do your tiling may not be great when considered in terms of the many years of use you will get from the tiles, especially if you have to buy or hire equipment to do the job yourself. But beware, not all general decorators are necessarily good at tiling. Ask to see other tiling work they have done and ask your tile supplier if they have a list of recommended tilers in your area.

There are whole books devoted to tiling, covering every minute detail, and some decisions you will have to make are a matter of personal taste. The following is intended only as a general guide, not a gospel. Remember to read carefully the manufacturer's instructions on the packages of adhesives, grouts and sealants, and above all, remember the golden rule of tiling which is to keep everything meticulously clean and orderly as work progresses. Many of the practices for fixing wall tiles are the same as for floor tiles, so read both sections even if you are only applying tiles to the walls. You will find that instructions for walls quite often say 'As for floors . . .'

For both floor and wall tiling there are two golden rules. The first is to satisfy yourself that the tiles you have chosen are suitable for the job you intend them to do. A reputable supplier should be happy to discuss this with you in detail, and also the method and materials for applying them. The second rule is not to rush. Take plenty of time to plan and prepare your walls and floors, and then to fix the tiles themselves. Archaeologists dig up tiles that have survived centuries; your tiles will be with you for a long time, and it is well worth taking the time to get them right.

ABOVE *If you are planning a patterned floor or border, lay the tiles out loose before fixing them in place, to be sure of getting it right.*

OPPOSITE *A floor or wall of coloured tiles need not be confined to the classic two-colour chequerboard effect. Several colours mixed together can be stunning, but plan carefully.*

FLOORS

EQUIPMENT

FOR PREPARING THE SUBFLOOR, YOU MAY NEED VARIOUS concreting and woodworking tools, depending on the extent of the work. For planning the layout of tiles on the floor you need a tape measure, string, chalk and a right angle. Fixing requires an adhesive trowel with a 6 mm (1/4 in) notched edge, and a spirit-level. You will need cutting tools (the heavier types can be hired) for fitting tiles around the edges of the room, and a selection of cloths, sponges and a paintbrush or roller, all clean and new, for sealing unglazed tiles. For grouting you need a hard implement like a piece of curved or bent narrow-gauge metal piping, to press the grout into the gaps between tiles. And finally you need a polisher or cloths for waxing, and a broom and mop for maintaining your newly tiled floor, as appropriate for the material from which the tiles are made.

THE SUBFLOOR

WHATEVER IT IS MADE OF, THE FLOOR TO BE TILED SHOULD BE level, clean, dry and free of dust, grease and loose material. This applies also to walls to be tiled.

TYPES OF SUBFLOOR

CONCRETE SCREED. When tiling on to a new screed, a period of at least one week for each 25 mm (1 in) depth of concrete should have elapsed to ensure that the screed is completely dry. Further time may be necessary depending on the weather and site conditions.

TIMBER. Suspended timber floors should be rigid and stable enough to bear the additional load. The floor can be stiffened by fixing 12 or 18 mm (1/2 or 3/4 in) 'shuttering' plywood over the existing timber, screwed down at 15 cm (6 in) intervals. Prime the surface of the plywood with a coat of an aqueous-based co-polymer (a general purpose primer and flexible compound which helps the adhesive cope with a certain amount of vibration). Do this before fixing the tiles, or add it to your adhesive. This will increase its durability and flexibility, and is also useful for laying tiles on a work surface in a kitchen.

EXISTING CERAMIC TILES. Ideally, old tiles should be lifted; but if they are completely sound, the new tiles can be laid over them. Existing glazed surfaces should be roughened with a sander to provide a key for the adhesive.

OTHER TYPES OF EXISTING FLOORING. Remove vinyl, cork and linoleum floors. Any bitumen left on the subfloor should be covered with a levelling compound such as a strong cement used in a thick, liquid state.

BUMPS AND HOLLOWS. Irregularities on the floor to be tiled can be taken up with a thick-bed adhesive during tiling. If the floor is very uneven, with undulations of more than 12 mm (1/2 in), a levelling compound should be used.

DAMP. Prior to fixing tiles, damp of any kind should be cured and, if necessary, a damp-proof membrane should be installed.

FINISHED FLOOR LEVEL. This will be raised by the thickness of the tiles to be laid, plus an adhesive bed of approximately 6 mm (1/4 in). Any resulting discrepancy with the levels of adjoining rooms may be resolved by installing timber threshold strips.

Planning

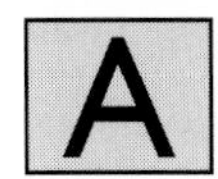
LLOW AMPLE TIME TO STUDY THE ROOM AND PLAN YOUR work. Tiling usually begins in the corner furthest from the door. If there is more than one door, you will need to decide which is the primary entrance. The following is a method for finding the exact position in which the first tile is to be laid, in a room of regular shape.

Step 1

Mark the centre line of the room from the wall of the primary entrance to the far end. To do this, measure and mark the midpoints of the wall with the door and the opposite wall. Join these two points with tight string and mark with a straight chalk line.

Step 2

Find the midpoint of the room by measuring halfway along the centre line. From this point, lay tiles on the floor along the centre line and going away from the door, without fixing them and allowing for the gap where the grouting will go (usually 5-10 mm ($\frac{1}{4}$-$\frac{1}{2}$ in), depending on the size of the tiles). This will show you where the last whole tile on this axis will come. Using string or chalk, mark a line across the room at a right angle to the centre line, along the outer edge of this last tile.

Step 3

Lay tiles along this new line (loose, allowing gaps for grouting, as before) towards the corner of the room. This will show you the position of the last whole tile on this axis, which is in fact where your very first whole tile should be laid on adhesive.

Step 4

As a check on the position of this first tile, mark a third line on the floor, along the outer edge of the tile, and make sure that you have a right angle and that this new line is parallel with the room's centre line. You can do this by measuring between the two lines at each end of the room and checking the measurements are identical.

* * *

IRREGULARLY SHAPED ROOM. With an irregularly shaped room, proceed as before after making the following adjustment. Create a regular shape with a right angle in each corner, within the room, using measuring tape, string and chalk, by lopping off triangles and any other shapes which distort the rectangle. When you have laid the tiles loosely on the floor (Steps 2 and 3 above), stand in the doorway and move around the room, to check that the tiles create the right visual lines for the room. It is no good being technically correct if the lines of tiles appear visually to be going off at an angle.

LAYING SQUARE TILES DIAGONALLY. When you have found the midpoint of the room (Step 2 above), lay two square tiles diagonally on the centre line so that a corner of each is almost on the midpoint, and the centre line also goes through the diagonally opposite corner of each. Remember to leave space for grouting between the tiles as before. Mark a line from wall to wall along the edge of the tile furthest from the door, and proceed as above to find the position of the last whole tile, which will be the first to be fixed down. Make the same checks on your right angles and parallel lines as you would if laying the tiles straight.

Fixing

P RIOR TO FIXING, CHECK THAT THE TILES ARE CLEAN, DRY AND generally of good quality. Shuffle the batch to ensure that any colour and tonal variations are well mixed. Do this with wall tiles too, if and when you do your wall tiling. Floor tiles are usually fixed with a thick-bed adhesive using the solid bed method as follows:

Step 1

Using an adhesive trowel, spread an area of around one square metre (10 square yards) to a depth of approximately 4 mm ($\frac{1}{6}$ in). Hand-made floor tiles may be irregular or even bowed and may therefore need a greater depth of adhesive in order to accommodate tiles of different thicknesses and level out potential irregularities in the finished floor surface. Some slate tiles vary considerably in thickness and are best laid on a sand and cement screed mixed in the ratio 3:1. Glazed insets used with terracotta floor tiles can be laid 2 or 3 mm ($\frac{1}{12}$-$\frac{1}{10}$ in) below the terracotta in order to minimize wear on the glaze.

Step 2

Butter the back of the first tile to be laid with a further 2 mm ($\frac{1}{12}$ in) of adhesive, and place it in the starting position. When laying thin, machine-made or quarry tiles on to an even surface it may be possible to use a thin bed (3 mm/$\frac{1}{10}$ in) of adhesive, without having to butter the back of the tile.

Step 3

Push the tile home firmly with a slight twisting movement, to spread the adhesive and force out any air beneath the tile. Place the next tile adjacent to the first, allowing for the width of the grout joint, and repeat this sideways and along, until the square metre is filled. All the tiles should be soundly bedded in the adhesive without any space for air underneath. Be careful not to allow any adhesive on to the surface of the tiles. If any appears, wipe it away immediately with a clean damp cloth. Check the floor with a spirit-level and a straight timber batten as you progress.

Step 4

When the square metre has been tiled, prepare another similar area and tile it, and continue in this way until you have covered the floor with whole tiles. The remaining perimeter area should now be prepared and tiled with cut tiles (see Cutting, below).

Cutting

IT IS POSSIBLE TO CUT SOME THIN, MACHINE-MADE FLOOR TILES using no more than a simple scoring and snapping tool. Mark the surface of the tile where it is to be cut, and firmly score down this line with the tungsten carbide wheel. Use the jaws of the cutter to make a clean break.

Most floor tile types are too thick and tough to be cut in this way, requiring more robust and sophisticated tile-cutting equipment. In general, an angle grinder (hired, or bought from a DIY shop) will do the job. Simply mark the tile where you wish to cut it and run the cutting disc along the marked surface three or four times until the tile can be snapped. Rough cut edges should be smoothed and finished with a sanding block or, in severe cases, by rubbing the tiled edge with a broken piece or the edge of another tile.

If your tiles are particularly tough or thick, or if you are tiling an exceptionally large area, you could make your task easier and quicker by hiring a water-cooled diamond cutter. Alternatively, ask about a tile-cutting service at your supplier or local tile shop.

Sealing unglazed tiles

MOST UNGLAZED TILES NEED TO BE SEALED IN ORDER TO protect the porous surface, but exact methods and timing depend on the material from which the tiles are made. Whatever the material, it is vital to make sure that your tiles, and the room, are completely clean and dry before you begin the sealing process. If they are not, you may actually seal in dirt, dust and moisture which could impair the floor's final appearance. If your tiles are damp, stand them on end in a well-ventilated area to dry thoroughly before laying. This may not be possible, in which case lay them and keep them clean but do not seal them until they are completely dry. Likewise, all equipment must be kept immaculately clean and materials such as brushes and cloths should be new and previously unused. Use a good quality sealant, preferably linseed-oil based.

Sealing usually takes place in two stages. The first seal is applied once the tiles are fixed but before they are grouted. The second seal goes on over the whole floor after grouting, once the grout has hardened and is completely dry. With sufficient space and time, you could alternatively seal your tiles individually before laying them. Another vital point is that the sealant must penetrate the tile. If you use too much and allow the excess to lie on the surface, it will harden and spoil. All sealants take a few days to harden fully once they have penetrated the tile, so keep traffic on your new floor light for at least the first week.

The following is a guide for sealing the various different materials from which floor tiles are made.

TERRACOTTA. Before you begin sealing the floor, test some sample tiles for porosity, to give you some idea of how much sealant they absorb. Apply the first seal evenly with a clean paintbrush or paint roller (do not pour it on directly). Especially porous types of terracotta may need two or three coats of sealer at both first and second seal. Work the sealant into the surface so that the tile appears matt again within between two and ten minutes. If the tile is matt in under one minute, use the sealant a little more liberally. Within twenty minutes of applying the sealant, check that there is absolutely no surplus remaining on the surface of the tiles. This is extremely important. Wipe away any surplus with a clean dry cloth.

Once the grouting has dried completely, give the floor a second seal by repeating the sealing process. Take great care not to apply too much sealant as the tiles will now be less porous. As before, make sure that you remove any surplus sealant from the surface.

RECLAIMED TERRACOTTA. It is quite likely that these tiles will recently have been cleaned by the supplier so take extra care to make sure that they are thoroughly dry before sealing. They may also be extremely dense due to years of wear. Test some samples for porosity and if they do not readily accept the sealer, mix it 1:1 with white spirit. Otherwise, seal in the same way as new terracotta.

QUARRY TILES. It is not strictly necessary to seal these tiles but sealing will give the floor added protection and make maintenance easier. Mix a good quality linseed oil sealant with white spirit in the ratio 1:1 and give the tiles a single coat, worked well into their surface.

ENCAUSTIC TILES. Handle these with great care as they are susceptible to staining until they are sealed. Use white spirit and a clean abrasive pad to remove any marks and consider sealing the tiles individually before laying them, unless you can take extreme care to keep them clean during the process. The first seal should consist of an impregnator which blocks the pores of the material. Apply this with a clean brush, fluff-free cloth or lambswool paint roller. Two hours later add the second coat, and two hours later a third if necessary. Do not use more sealant than the tiles will absorb fully, remaining matt, and apply the second coat at right angles to the first. The second seal should be two coats of an acrylic polymer to give the tiles a protective coating on the surface, applied in the same way as the first.

MARBLE. Marbles vary hugely in finish – some are already highly polished. Consult your supplier about treatment appropriate to his tiles. Rough marble can be sealed with two coats of an appropriate sealant once the grouting is hard and dry and the floor has been washed clean and allowed to dry. The second coat should be added before the first is fully dry.

SLATE. Like marble, slate can be sealed in one continuous process. Pour an acrylic polymer that will give a protective surface coating on to the tiles and spread in one direction only with a fluff-free cloth or paint roller. Do not rub or buff. Allow it to dry for about twenty minutes then apply a second coat at right angles to the first, to ensure an even coverage.

LIMESTONE. Special limestone sealant is available, as unsealed limestone is susceptible to staining. Read the manufacturer's instructions carefully. The first coat goes on before grouting, applied thinly and worked into the tiles with a new paint brush. After about an hour it should have dried sufficiently to allow light traffic and grouting. When the grout is completely dry add a second coat, covering the grouting as well, and leave to cure for about forty-eight hours.

GROUTING

THIS IS THE PROCESS OF FILLING THE SPACES BETWEEN INDIVIDUAL tiles and thus completing the floor surface. The purpose of grouting is to strengthen the overall floor, and in order to do this the grout must support the edges of all the tiles fully. This means filling the gaps completely, level or almost level with the top of the tile, not sunk down leaving vulnerable edges exposed.

Depending on site conditions, the fixed tiles should be left for between twelve and twenty-four hours, to allow the adhesive bedding to dry out. The joints should be clean, dry and free of loose debris. Check the manufacturer's recommendations to ensure that your grout is suitable for the type of tiles and situation. For example, sandstone grout is useful for most domestic tiling and with more rustic, textured floor tiles. Grey grout is good for glazed tiles where you will want a more precise grout joint, but you should use it with care with unglazed tiles as the fine particle composition carries a higher risk of staining the tiles. Sand and cement mixed 3:2 can make an effective grout; use fine washed sand for a sandstone appearance or a sharp sand for a more rustic look. Joints of more than 10 mm (½ in) will need a wide-joint grouting compound.

In general it is better to fill the joints by a pointing process, pressing it neatly into the gaps, rather than a spreading or slurry technique. Porous, unglazed tiles which could be stained by excess grout should certainly be grouted by pointing, and their surface must be kept clean at all times by being wiped with a damp, clean sponge. Push the grout well down into the joints to prevent air pockets and help the grout set hard. Then smooth off the joints with a hard implement slightly narrower than the grout – a small piece of narrow, bent metal pipe (not copper) would be suitable.

As soon as the floor is finished and the grout has completely dried, the tiles should be thoroughly brushed to remove any stains or fine dust particles. If necessary, wipe them with a damp sponge and leave them to dry.

MAINTENANCE

ALL FLOORS BENEFIT FROM SOME MAINTENANCE, IF ONLY SWEEPing and wiping over with a mop and light detergent once in a while. Unglazed tiles, particularly terracotta which improves with wear, need a little more maintenance.

Polish terracotta with a good quality wax polish in paste form, use it sparingly and finish it with plenty of effort to build up a hard protective finish. A first polish should be given to the finished floor just a few hours after the final seal, and two more in quick succession; then once a week for the following four or six weeks to build up a hard-wearing surface.

Some types of pre-sealed terracotta require a maintenance kit which should be available from the supplier and used as instructed. Quarry tiles can be polished with a good quality wax. Encaustic, marble and slate tiles should be maintained with a low-alkaline detergent diluted with warm water. Rinse the mop regularly in the solution to remove dirt as it accumulates. If you want a shine on slate tiles, finish with wax polish.

WALLS

Equipment

You will need the same equipment as for floor tiling, with the exception of sealing equipment (if you are using glazed tiles). In addition you will want timber battens measuring about 50 x 25 mm (2 x 1 in) and a plumb line, for planning the layout of your tiles, a 3 mm (1/10 in) tiling trowel and adhesive spreader for fixing, tile pincers or a hacksaw fitted with a multi-directional tungsten carbide blade for cutting curves and other complex shapes, and a rubber squeegee or float (not metal which could damage the glaze) for grouting.

Wall preparation

Tiles can be fixed to most walls. As with floors, the surface must be properly prepared and be even, sound, clean, dry and free of dust, grease or any loose material.

Unrendered and rendered walls. Concrete block and other types of unrendered wall should be tiled using a thick-bed adhesive to allow for the uneven surface of the wall. Tiles can be fixed directly on to sand and cement render (4:1 using Portland cement) which in fact is an ideal base. Newly built or rendered walls will need at least fourteen days to dry out thoroughly before tiling can begin.

Plastered, papered and painted walls. Loose plaster must be made good, and new plaster should have been completed at least four weeks before tiling and be primed with a plaster primer. Old wallpaper should be stripped. Sand gloss paint to provide a key for the adhesive, and remove any peeling gloss or emulsion paint.

Timber. Seal walls of timber or timber composite construction with a bonding agent, and allow it to dry before tiling.

Existing tiled walls. Old tiles do not have to be removed, so long as they are secure and the wall underneath is sound. If necessary, sand to provide a key for adhesive.

Bumps, hollows and damp. Serious defects in the wall surface should be smoothed and made good. The wall must be completely dry – any underlying damp problem must be solved before fixing tiles to it.

Planning

Take plenty of time to do this. The most important part of planning wall tiling is to decide where the focal point is on the wall or in the room and to ensure that your view won't be obscured by objects placed in front of the tiles. This is the place where you want an immaculate row of whole tiles, not a strip of tile pieces that have had to be cut to fit leftover space. On a kitchen wall this will be either immediately above the work surface, or possibly just below the wall cupboards, where clutter on the work surface will not hide the tiles. In a bathroom, the place where you feel it is most important to have whole tiles rather than pieces may be just below the ceiling, or above the skirting, or, most likely, along the edge of the bath. The choice is entirely up to you and your personal taste. You will have to live with it, so take your time to get it right.

Once you have decided on the focal point, proceed in much the same way as for floor tiling, using batons, measure and plumb line to work out where your first tile should be fixed. Use a spirit-level regularly to check the tiles are straight, and remember to leave room for the grouting, both when planning and fixing. Nail battens at right angles to each other creating a corner where the first tile in the bottom whole row of tiles will rest, and don't remove the battens until the adhesive is completely dry.

FIXING

U SE NON-SLIP, WATER-RESISTANT CERAMIC WALL TILE ADHESIVE suitable for intermittently wet areas such as work surfaces, splashbacks, bathrooms and showers. An exception to this is a situation where the tiles may be completely or continuously immersed in water (the lining of a bath, for example). In this case, use a specialist adhesive of a type which is also suitable for fixing tiles on exterior walls and floors.

With an adhesive spreader, apply a thin (3 mm/$^1/_{10}$ in) layer of adhesive to an area on the wall of about one square metre (10 square yards). With the notched side of the spreader, comb the area to create a ribbed surface of adhesive which will accept the tiles. Position the first tile carefully in the corner formed by the timber battens and push the tile home firmly with a slight twisting movement to spread the adhesive and ensure that there are no pockets of air trapped behind. Fill the first square metre, then prepare another, and so on, as with floors.

Some machine-made tiles have built-in lugs to predetermine the space between tiles. With the majority of glazed tiles, however, you will have to use spacers or small pieces of cardboard the thickness of the gap you wish to leave for the grouting. Place these between the tiles as you bed them on to the adhesive. Once you have completed the wall using whole tiles, remove the timber battens and tile around the perimeter with cut tiles.

CUTTING

A S WITH FLOORS, THE TOUGHER THE TILE, THE MORE sophisticated the tile cutting equipment needs to be. Machine-made wall tiles can usually be cut easily with a hand-held cutter. Curves and other more complex shapes can be cut out with a pair of tile pincers or a hacksaw fitted with a multi-directional tungsten carbide blade. Rustic and hand-made tiles are denser and need to be cut with an angle-grinder, as do stoneware tiles. Mitres or angles across the edges or faces of tiles need a precise corner finish, best provided by a diamond cutter.

GROUTING

G ROUTING WALLS IS MUCH THE SAME AS FOR FLOORS, EXCEPT that a spreading technique is better than a pointing one for filling the joints. Use a water-resistant grout and check manufacturer's instructions to ensure that your grout is suitable. In places like showers, use an epoxy grout and in any damp place finish the grout flush with the glazed surface to avoid water seeping behind the tiles. Apply the grout with a rubber squeegee or grouting trowel, working it backwards and forwards across the tiles until all the joints are completely filled. As the grout is drying, wipe any excess smears from the glazed surface with a clean sponge. The space or join between tiles and a bath, basin, sink or worktop should be sealed with a silicone sealant to prevent moisture penetrating behind the tiles.

TILE SOURCE LIST

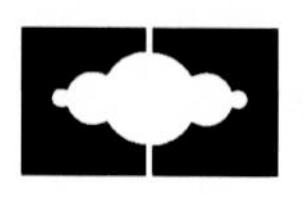

UK TILE SOURCES

FIRED EARTH SHOWROOMS

102 Portland Road, London W11 4LX
071 221 4825

21 Battersea Square, London SW11 3RA
071 924 2272

187 New Kings Road, London SW6 4SW
071 736 5987

3 Saracen Street, Bath BA1 5BR
(0225) 442594

69 Calverley Road, Tunbridge Wells TN1 2UY
(0892) 540 220

2 Church Street, Wilmslow SK9 1AU
(0625) 548 048

Middle Aston, Oxford OX5 3PX
(0869) 40724

60 Holywell Hill, St Albans AL1 1BX
(0727) 55407

57 Moulsham Street, Chelmsford CM2 0JA
(0245) 494 684

13 Woodstock Road, Oxford OX2 6HA
(0865) 514 549

25 Clarence Parade, Cheltenham GL50 3PA
(0242) 251 455

21 Winchester Street, Salisbury SP1 1HB
(0722) 414 554

36 Lower Bridge Street, Chester CH1 1RS
(0244) 348 084

114 Regent Street, Leamington Spa CV32 4NR
(0926) 886 125

14 King Street, Nottingham NG1 2AY
(0602) 476 534

2 Chapel Street, Guildford GU1 3UH
(0483) 300 052

15 Burgate, Canterbury CT1 2HG

ANTA
141 Portland Road, Clarendon Cross,
London W11 4LR
071 229 5077

MAGGIE ANGUS BERKOWITZ
21-23 Park Road, Milnthorpe, Cumbria
LA7 7AD
(0539) 563 970

CASA CERAMICS
24 Fishergate Hill, Preston, Lancashire PR1 8JB
(0772) 201 443

CASBAH TILES
20 Wellington Lane, Montpelier, Bristol,
Avon BS6 5PY
(0272) 427 318

CASTELNAU TILES
175 Church Road, London SW13 9HR
081 741 2452

CORRES MEXICAN TILES
15 Ewer Street, London SE1 0NR
071 261 0941

CRITERION TILES
196 Wandsworth Bridge Road, London
SW6 2UF
071 736 9610

THE DECORATIVE TILE WORKS
(Bespoke Tile Makers)
Jackfield Tile Museum, Ironbridge, Telford,
Shropshire TF8 7AW
(0952) 884 124

DOMUS TILES
33 Parkgate Road, London SW11 4NP
071 223 5555

EDGAR UDNY & CO.
314 Balham High Road, London SW17 7AA
081 767 8181

ELON
66 Fulham Road, London SW3 6HH
071 584 8966

JONES'S TILES
Orleton Manor, Ludlow, Shropshire SY8 4HR
(0568) 856 66

KENNETH CLARK CERAMICS
The North Wing, Southover Grange,
Southover Road, Lewes,
East Sussex BN7 1TP
(0273) 476761

LAURA ASHLEY
150 Bath Road, Maidenhead,
Berkshire SL6 4YS
(0628) 39151

THE LIFE ENHANCING TILE COMPANY
Unit 4a, Alliance House,
14/28 St Mary's Road, Portsmouth,
Hampshire PO1 5PH
(0705) 862 709

MARLBOROUGH CERAMIC TILES
Elcot Lane, Marlborough,
Wiltshire SN8 2AY
(0672) 512 422

MEGA CERAMICS
8 Station Yard, Oxford Road, Adderbury,
Oxon OX17 3HP
(0295) 812288

THE ORIGINAL TILE COMPANY
23A Howe Street, Edinburgh Lothian EH3 6TF
031 556 2013

PARIS CERAMICS
583 Kings Road, London SW6 2EH
071 371 7778

PORTUGUESE VINTAGE TILES
Ring (0285) 644 833 for stockists

PURBECK DECORATIVE TILE CO.
20 Smugglers Way, London SW18 1EQ
081 871 0570

THE REJECT TILE SHOP
178 Wandsworth Bridge Road, London
SW6 2UQ
071 731 6098

TERRA FIRMA TILES
70 Chalk Farm Road, London NW1 8AN
071 485 7227

THE TILE GALLERY
50A North Street, Sudbury,
Suffolk CD10 6RE
(0787) 706 29

TILES OF STOW
Langston Priory Workshops, Station Road,
Kingham, Oxon OX7 6UP
(0608) 6589 51

THE TILE SHOP
6 Chamberlayne Road, London NW6 3JD
081 968 9497

VILLEROY & BOSCH
27 Victoria Street, Windsor, Berkshire
SL4 1HE
(0753) 857778

WELLINGTON TILE COMPANY
Tonedale Industrial Estate, Milverton Road,
Wellington, Somerset TA21 0AZ
(0823) 667 242

BRONWYN WILLIAMS-ELLIS
The Basement, Cleveland Bridge Gallery,
8 Cleveland Place East, Bath, Avon BA1 5DJ
(0225) 447 885

WHICHFORD POTTERY
Shipston-on-Stour, Warwickshire CV36 5PG
(0608) 84416

USA TILE SOURCE

ANTIQUE FLOORS
Dallas TX
214 760 9330

DESIGN TILE
Tyson Corners VA
703 734 8211

FACINGS OF AMERICA
Phoenix AZ
602 955 9217

FIRED EARTH
ABC
888 Broadway
New York NY
212 473 3000

FLORIDA TILE
Salt Lake City UT
801 485 2900

HAMILTON PARKER
Columbus OH
614 221 6593

NEVADA TILE
Las Vegas NV
702 388 7000

THE TILE COLLECTION
Pittsburgh PA
512 727 7279

TILECRAFT
San Francisco CA
415 552 1913

TILE SOURCE
Overland Park KS
913 345 8453

VIRGINIA TILE
Troy MI
313 649 4422

SHELLY TILE
New York NY
212 832 2255

CHARLES TILES
Baltimore MD
410 332 1500

SUNDERLAND BROTHERS
Omaha NE
402 339 2220

SUNNY MCLEAN
Miami FL
305 573 5943

TILE CONTRACTORS SUPPLY
Nashville TN
615 269 9669

TILES, A REFINED COLLECTION
Boston MA
617 537 0400

TRADITIONS IN TILES
Atlanta GA
404 239 9186

UNITED TILE
Portland OR
503 231 4959

DEDICATION

FOR
Archie and Harriet

ACKNOWLEDGEMENTS

I WOULD LIKE TO THANK EVERYONE AT FIRED EARTH for their enthusiasm for this book and for their generous supply of information, but especially Nicholas Kneale, Philippa Humphrey, Alan Berks (author of *The Tilers Art* guides which are the basis of chapter twelve), Robert Tate and Robin Buchanan.

I WOULD ALSO LIKE TO THANK THE FOLLOWING PEOPLE and organizations for their generous help: Bal (Building Adhesives) Ltd; Juliet Beaumont; Maggie Angus Berkowitz; Felicity Bryan; Georgina Cardew; Professor John Carswell, Sotheby's; Kenneth Clark; John de Falbe; Paul Henry; Tony Herbert, Tiles and Architectural Ceramics Society; Mrs David Jackson; Sebastian John; Bobby Jones; Rachel King; Hans van Lemmen, Leeds Metropolitan University; Barbara and John Milligan; Sarah Riddick; John Rumsby, Tolson Memorial Museum, Huddersfield; Mrs Max Selka; William Selka; Simon Sheard; Helen Sudell; Susan Taylor; Odette Wells; Westminster Abbey press office; Bronwyn Williams-Ellis; Fenella Willis; Whichford Pottery; Shona Wood.

Elizabeth Hilliard

INDEX

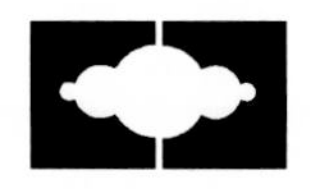

Illustrations have page numbers in italic.

Picture Acknowledgements

All Fired Earth except:

Laurie Black p. 72, 86, 91, 92, 94, 97, 99, 100 top, 101, 104, 106, 109, 110

Bridgeman Art Library p. 6

Copyright British Museum p. 16, 19, 20, 21, 24

Dean & Chapter of Westminster Abbey p. 23

Kenneth Clark Ceramics p. 58, 66 top left, 67

Richard Dennis Publications p. 17 top, 46, 49, 50, 52, 54, 55, 56, 57, 60, 63

English Heritage p. 25

Harrods p. 53

Portuguese National Tourist Office p. 34

The Royal Borough of Kensington & Chelsea p. 14

Tiles of Stow p. 108

Bronwyn Williams Ellis p. 62, 65, 66 below

Elizabeth Whiting & Associates: Andreas von Eisiedel p. 41

Brian Harrison p. 75, Graham Henderson p. 2 & back Jacket

Rodney Hyett p. 78 left, 87, 90, 102, 112, 119

Neil Lorimer p. 74, 77, 95, 96, 120 & back Jacket

Peter Woloszynski p. 117

Mount Laurel Library
100 Walt Whitman Avenue
Mt. Laurel, N.J. 08054-9539
(609) 234-7319